Competency-based Comprehensive Manual of
Practical and Clinical
BIOCHEMISTRY

Competency-based Comprehensive Manual of

Practical and Clinical BIOCHEMISTRY

As per Revised NMC Curriculum

Second Edition

Ashish Sharma MBBS MD
Professor and Head
Department of Biochemistry
Geetanjali Medical College and Hospital
Udaipur, Rajasthan, India

Anita Sharma MBBS MD
Professor and Head
Department of Biochemistry
Lab-in-Charge Clinical Biochemistry
Himalayan Institute of Medical Sciences
Dehradun, Uttarakhand, India

JAYPEE BROTHERS MEDICAL PUBLISHERS
The Health Sciences Publisher
New Delhi | London

 **Jaypee Brothers Medical Publishers (P) Ltd**

Headquarters
Jaypee Brothers Medical Publishers (P) Ltd
EMCA House, 23/23-B
Ansari Road, Daryaganj
New Delhi 110 002, India
Landline: +91-11-23272143, +91-11-23272703
+91-11-23282021, +91-11-23245672
Email: jaypee@jaypeebrothers.com

Corporate Office
Jaypee Brothers Medical Publishers (P) Ltd
4838/24, Ansari Road, Daryaganj
New Delhi 110 002, India
Phone: +91-11-43574357
Fax: +91-11-43574314
Email: jaypee@jaypeebrothers.com

Overseas Office
J.P. Medical Ltd
83 Victoria Street, London
SW1H 0HW (UK)
Phone: +44 20 3170 8910
Fax: +44 (0)20 3008 6180
Email: info@jpmedpub.com

Website: www.jaypeebrothers.com
Website: www.jaypeedigital.com

© 2021, Jaypee Brothers Medical Publishers

The views and opinions expressed in this book are solely those of the original contributor(s)/author(s) and do not necessarily represent those of editor(s) of the book.

All rights reserved. No part of this publication may be reproduced, stored or transmitted in any form or by any means, electronic, mechanical, photocopying, recording or otherwise, without the prior permission in writing of the publishers.

All brand names and product names used in this book are trade names, service marks, trademarks or registered trademarks of their respective owners. The publisher is not associated with any product or vendor mentioned in this book.

Medical knowledge and practice change constantly. This book is designed to provide accurate, authoritative information about the subject matter in question. However, readers are advised to check the most current information available on procedures included and check information from the manufacturer of each product to be administered, to verify the recommended dose, formula, method and duration of administration, adverse effects and contraindications. It is the responsibility of the practitioner to take all appropriate safety precautions. Neither the publisher nor the author(s)/editor(s) assume any liability for any injury and/or damage to persons or property arising from or related to use of material in this book.

This book is sold on the understanding that the publisher is not engaged in providing professional medical services. If such advice or services are required, the services of a competent medical professional should be sought.

Every effort has been made where necessary to contact holders of copyright to obtain permission to reproduce copyright material. If any have been inadvertently overlooked, the publisher will be pleased to make the necessary arrangements at the first opportunity. The **CD/DVD-ROM** (if any) provided in the sealed envelope with this book is complimentary and free of cost. **Not meant for sale**.

Inquiries for bulk sales may be solicited at: jaypee@jaypeebrothers.com

Competency-based Comprehensive Manual of Practical and Clinical Biochemistry

First Edition: 2020

Second Edition: **2021**

ISBN: 978-93-5270-055-4

Printed at: Sterling Graphics Pvt. Ltd.

Dedicated to

The first teachers of our life, who patiently taught us everything….our parents.

Preface to the Second Edition

National Medical Commission of India has recently revised MBBS curriculum and we have tried to incorporate all the relevant exercise and modifications in this new edition.

We are overwhelmed with convivial suggestions as well as constructive criticism received from medical fraternity across the country. We have tried to incorporate most of suggestions and move forward with deep gratitude for all those who helped us to improve this book.

Ashish Sharma

Highlights of the book are:

- Attitude, ethics and communication modules: As per NMC, AETCOM are essential components for competency-based medical teaching, so in this edition we add four AETCOM modules
- Competency-based assessment
- Early clinical exposure and self-directed learning
- Spotting for practical.

Anita Sharma

We hope that this edition will be more constructive for students as well as for faculty members. We welcome all our readers for submission of their ideas and feedback for further development of this competency-based practical manual.

Preface to the First Edition

In the ever-changing and challenging fields of medicine, biochemistry is the mainstay in the basic of development, diagnosis and therapeutics of various human diseases. That is why it is imperative to remain abreast of the latest development.

After the initiative of World Federation for Medical Education (WFME), it is recommended that competency-based medical education should be an integral part of curriculum of undergraduate medical students. Keeping this idea in mind, a complete and comprehensive practical and clinical biochemistry book based on GMR (Graduate Medical Regulations)–2019 is developed.

This book contains five sections, each section is further divided into chapters inclusive of all core components of practical and clinical biochemistry. Each chapter explains specific competencies, learning objectives, level of learning and learning domain involved in clinical and practical biochemistry.

The highlights of the manual are:

- **Basic laboratory principles:** It would enable students to know about apparatus and equipment used in the laboratory, good and safe laboratory practice, and waste disposal systems in laboratory, quality assurance and quality control in clinical biochemistry.
- Quantitative experiments and their interpretation with early clinical exposure including case studies and viva voce questions, which provide adequate knowledge and skill to interpret and perform these laboratory tests independently.
- **Group experiments, demonstration, self-directed learning, attitude and communication skill development exercise, group discussions** are planned in such a lucid manner that encourage students to participate and to instill in them various advanced clinical and molecular biology experiments.

We hope this effort would enable the students to focus more in developing their knowledge of clinical interpretation rather than writing practical files and hence will upgrade their skills in practical and clinical biochemistry.

Ashish Sharma
Anita Sharma

Reviewers List

Asha Khubchandani
Professor and Head
Department of Biochemistry
BJ Medical College
Ahmedabad, Gujarat, India

Debabrata Dash
Professor and Head
Department of Biochemistry
Institute of Medical Sciences
Banaras Hindu University
Varanasi, Uttar Pradesh, India

Hitesh Shah
Professor and Head
Department of Biochemistry
Pramukhswami Medical College
Karamsad, Gujarat, India

Jitendra Ahuja
Associate Professor
Department of Biochemistry
RUHS College of Medical Sciences
Jaipur, Rajasthan, India

Kiran Chauhan
Professor
Department of Biochemistry
GMERS Medical College
Gandhinagar, Gujarat, India

Mahendra Kumar Gajanan Dhabe
Associate Professor
Department of Biochemistry
Swami Ramanand Teerth Rural Medical College
Ambajogai, Beed, Maharashtra, India

Manisha Arora
Chair Professor Biochemistry
Medical University of the Americas
Charlestown, Nevis, West Indies

Manisha Nathani
Additional Professor
Department of Biochemistry
All India Institute of Medical Sciences
Rishikesh, Uttarakhand, India

Sanjeev Singh
Professor and Head
Department of Biochemistry
Gajra Raja Medical College
Gwalior, Madhya Pradesh, India

Shuchi Goyal
Professor and Head
Department of Biochemistry
Ravindra Nath Tagore Medical College
Udaipur, Rajasthan, India

Vilas U Chavan
Professor and Head
Department of Biochemistry
Surat Municipal Institute of Medical Education and Research (SMIMER)
Surat, Gujarat, India

Advisory Committee

Ashish Jadhav
Additional Professor
Department of Biochemistry
All India Institute of Medical Science
Bhopal, Madhya Pradesh, India

Gaurav Modi
Associate Professor
Department of Biochemistry
GMERS Medical College
Vadnagar, Gujarat, India

Neha Sharma
Associate Professor
Department of Biochemistry
Geetanjali Medical College and Hospital
Udaipur, Rajasthan, India

Nita Sahi
Associate Professor
Department of Biochemistry
Pacific Medical College and Hospital
Udaipur, Rajasthan, India

Shrawan Kumar
Professor and Head
Department of Biochemistry
Pandit Deendayal Upadhyaya Medical College
Churu, Rajasthan, India

Sohil Takodara
Associate Professor
Department of Biochemistry
Geetanjali Medical College and Hospital
Udaipur, Rajasthan, India

Suresh Gautam
Assistant Professor
Department of Biochemistry
Ananta Medical College
Udaipur, Rajasthan, India

Uday Vachhani
Associate Professor
Department of Biochemistry
GMERS Medical College and Hospital
Himmatnagar, Gujarat, India

Ummed Solanki
Professor and Head
Department of Biochemistry
Jhalawar Medical College
Jhalawar, Rajasthan, India

Contents

UNIT 1: INTRODUCTION OF CLINICAL BIOCHEMISTRY

1.1.	Laboratory Apparatus and Equipment, Good and Safe Laboratory Practice, and Waste Disposal Systems in Laboratory	3

UNIT 2: QUALITATIVE EXPERIMENTS AND THEIR CLINICAL APPLICATIONS

2.1.	Analysis of Carbohydrates	13
2.2.	Analysis of Proteins	28
2.3.	Analysis of Physical and Chemical Composition of Physiological Urine	48
2.4.	Identify, Perform and Interpret Pathological Urine Analysis and Correlate it with Pathological States	56

UNIT 3: QUANTITATIVE EXPERIMENTS AND THEIR CLINICAL INTERPRETATION

3.1.	Principle of Colorimetry	67
3.2.	Principle of Spectrophotometry	72
3.3.	Estimation of Blood Glucose	78
3.4.	Glucose Tolerance Test and Glycated Hemoglobin	86
3.5.	Liver Function Test	92
3.6.	Kidney Function Test	113
3.7.	Lipid Profile (Atherogenic Profile)	124
3.8.	Estimation of Serum Calcium and Serum Phosphorus	134

UNIT 4: SELF-DIRECTED LEARNING EXERCISES

4.1.	pH Meter	141
4.2.	Water Homeostasis and Estimation of Na^+ and K^+ with ISE Analyzer	144
4.3.	Arterial Blood Gas Analyzer	147
4.4.	Chromatography	150
4.5.	Electrophoresis	156
4.6.	Enzyme-linked Immunosorbent Assay	160
4.7.	Antigen-Antibody Interaction (Immunodiffusion)	164
4.8.	Quality Control in Clinical Laboratory	170
4.9.	DNA Isolation from Blood and Tissue	175

UNIT 5: EARLY CLINICAL EXPOSURE EXERCISES AND REFLECTIVE WRITING

5.1.	Analysis of Cerebrospinal Fluid	181
5.2.	Thyroid Function Test	186
5.3.	Pancreatic Function Tests	192
5.4.	Disorders of Acid-Base Balance	197

UNIT 6: ATTITUDE, ETHICS AND COMMUNICATION (AETCOM) MODULES

6.1.	Introduction of Clinical Methods	205
6.2.	What does it Mean to be a Doctor?	207
6.3.	What does it Mean to be a Patient?	209
6.4.	The Doctor-Patient Relationship	211
6.5	The Foundations of Communications	215

UNIT 7: BIOCHEMICAL CALCULATIONS AND REFERENCE RANGE

7.1.	Preparations of Buffers and Solutions	221
7.2.	Reference Value of Various Biochemical Parameters Integration with Medicine	223

UNIT 8: PRACTICAL SPOTS IN BIOCHEMISTRY

8.1.	Practical Spots in Biochemistry	229

UNIT 9: COMPETENCY-BASED ASSESSMENT FOR PRACTICAL BIOCHEMISTRY

9.1.	Competency-based Assessment for Practical Biochemistry	235

Index   *243*

List of Experiments

S.No.	Experiment	Page No.	Date	Teacher Signature
Unit 1	**Introduction of Clinical Biochemistry**			
1.1	Laboratory Apparatus and Equipment, Good and Safe Laboratory Practice, and Waste Disposal Systems in Laboratory			
Unit 2	**Qualitative Experiments and their Clinical Applications**			
2.1	Analysis of Carbohydrates			
2.2	Analysis of Proteins/Amino Acid			
2.3	Physiological Urine Analysis			
2.4	Pathological Urine Analysis			
Unit 3	**Quantitative Experiments and their Clinical Interpretation**			
3.1	Principle of Colorimetry			
3.2	Principle of Spectrophotometry (com B.1.11.18)			
3.3	Blood Glucose Estimation			
	Case Study: Diabetes Mellitus			
3.4	Glucose Tolerance Test (GTT) and Glycated Hb (GHb)			
	Case Study: Gestational Diabetes			
3.5	Assessment of Liver Function Test (LFT)			
3.6	Estimation of Serum Bilirubin			
3.7	Estimation of Serum Transaminases SGOT/SGPT			
3.8	Estimation of Serum Alkaline Phosphatase			
	Case Study: Jaundice			
3.9	Estimation of Protein and A/G ratio			
	Case Study: Proteinuria/Nephrotic Syndrome			
3.10	Assessment of Kidney Function Test			
3.11	S. Creatinine and Creatinine Clearance			
3.12	Serum Urea and Urea Clearance			
	Case Study: Renal Failure			
3.13	Serum Uric Acid			
	Case Study: Gout			
3.14	Serum Lipid Profile			
3.15	Estimation of Total Cholesterol			
3.16	Estimation of Triglyceride			
3.17	Estimation of HDL Cholesterol			
	Case Study: Dyslipidemia and Myocardial infarction			
3.18	Estimation of Serum Calcium and Phosphorus			

S.No.	Experiment	Page No.	Date	Teacher Signature
Unit 4	**Self-directed Learning Exercises**			
4.1	pH Meter and Measurement of pH of Buffers			
4.2	Water Electrolytes Homeostasis and Estimation of Na^+ and K^+ with ISE Analyzer			
4.3	Blood Gas analysis by ABG analyzer			
4.4	Paper Chromatography and Thin Layer Chromatography (TLC)			
4.5	Protein Electrophoresis (PAGE)			
4.6	ELISA			
4.7	Immunodiffusion and Immunochemical Analysis			
4.8	Quality Control in Clinical Laboratory			
	Reflective Writing			
4.9	DNA Isolation from Blood and Tissue			
Unit 5	**Early Clinical Exposure Exercises and Reflection Writing**			
5.1	CSF Examination			
	Reflective Writing			
5.2	Thyroid Function Test			
	Reflective Writing			
5.3	Pancreatic Function Tests			
	Reflective Writing			
5.4	Disorders of Acid-Base Balance			
	Reflective Writing			
Unit 6	**Attitude, Ethics and Communication (AETCOM) Modules**			
6.1	Introduction of Clinical Methods			
6.2	What does it Mean to be a Doctor?			
6.3	What does it Mean to be a Patient?			
6.4	Doctor-Patient Relationship			
6.5	Foundation of Communication -1			
Unit 7	**Biochemical Calculations and Reference Range**			
7.1	Preparations of Buffer and Solutions			
7.2	Reference Value of Various Biochemical Parameters Integration with Medicine			
Unit 8	**Practical Spots in Biochemistry**			
Unit 9	**Competency-based Assessment**			

Competency-based MCI Curriculum for Biochemistry

S.No.	Experiment	Competency Covered	Early Clinical Exposure/ATCOM/ Integration (Vertical/Horizontal)/Certification
Unit 1	**Introduction of Clinical Chemistry**		
1.	Laboratory Apparatus and Equipments, Good and Safe Laboratory Practice, Waste Disposal Systems in Laboratory	BI. 11.1 and BI. 11.19	
Unit 2	**Qualitative Experiments and their Clinical Applications**		
2.	Analysis of General Carbohydrates	BI.3.1 and BI.3.8	BI.3.8 (Pathology and general medicine)
3.	Analysis of Proteins/Amino Acid	BI.5.4 and BI.5.5	BI.5.4 (Pediatrics) BI.5.5 (General medicine)
4.	Physiological Urine Analysis	BI.11.3 and BI.11.4	BI.11.4 (Physiology and general medicine) BI.11.4 (Perform level competency with certification)
5.	Pathological Urine Analysis	BI.11.4 and BI.11.20	BI.11.4 (Physiology and general medicine) PE.21.11 & PE.33.6 (Pediatrics) IM.11.13 (Internal medicine) BI.11.4 (Perform level competency with certification) BI.11.20 (Competency with certification)
Unit 3	**Quantitative Experiments and their Clinical Interpretation**		
6.	Principle of Colorimetry	BI.11.6	
7.	Principle of Spectrophotometry	BI.11.18	
8.	Blood Glucose Estimation	BI.3.8, BI.3.10 BI.11.21, IM.11.12	BI.3.8 (Pathology and general medicine) BI.3.10 (General medicine) BI.11.21 (Competency with certification) IM 11.12 (Pathology, general medicine: Perform level competency with certification)
	Case Study: Diabetes Mellitus		Early clinical exposure
9.	Glucose Tolerance Test (GTT) and Glycated Hb (GHb)	BI.3.8, BI.3.10	BI.3.8 (Pathology and general medicine) BI.3.10 (General medicine)
	Case Study: Gestational Diabetes		Early Clinical Exposure
10.	Assessment of Liver Function Test (LFT)	BI.6.13, BI.6.14 BI.11.17, PY.4.8	BI.6.13, BI.6.14 (Physiology, anatomy, pathology, general medicine) PY.4.8 (Physiology)

S.No.	Experiment	Competency Covered	Early Clinical Exposure/ATCOM/Integration (Vertical/Horizontal)/Certification
10.1.	Estimation of Serum Total Protein	Bl.11.8, Bl.11.17, Bl.11.21	Bl.11.8 (Perform level competency with certification) Bl.11.17 (General medicine, pathology) Bl.11.21 (Competency with certification)
10.2.	Estimation of Albumin and A/G Ratio	Bl.11.8, Bl.11.17, Bl.11.22	Bl.11.8 (Perform level competency with certification) Bl.11.17 (General medicine, pathology) Bl.11.22 (General medicine)
	Case Study: Proteinuria/Nephrotic Syndrome		Early clinical exposure
10.3.	Estimation of Serum Bilirubin	Bl.11.12, Bl.11.17	Bl.11.12 (Perform level competency) Bl.11.17 (General medicine, pathology)
10.4.	Estimation of Serum Transaminases SGOT/SGPT	Bl.2.2, Bl.11.13	Bl.11.13 (Perform level competency)
10.5.	Estimation of Serum Alkaline Phosphatase	Bl.11.14	Bl.11.14 (Perform level competency)
	Case Study: Jaundice		Early clinical exposure
11.	Assessment of Kidney Function Test	Bl.6.13, Bl.6.14, Bl.11.17, PY.7.8	Bl.6.13, Bl.6.14 (Physiology, anatomy, pathology, general medicine) Bl.11.17 (General medicine, pathology) PY.7.8 (Physiology)
11.1.	Serum Creatinine and Creatinine Clearance	Bl.11.7, Bl.11.22	Bl.11.7 (Perform level competency with certification) Bl.11.22 (General medicine)
11.2.	Serum Urea and Urea Clearance	Bl.11.21	Bl.11.21 (Competency with certification)
	Case Study: Renal Failure		Early clinical exposure
12.	Serum Uric Acid	Bl.11.17	Bl.11.17 (General medicine, Pathology)
	Case Study: Gout		Early clinical exposure
13.	Serum Lipid Profile	Bl.4.5/Bl.4.7 Bl.11.9, Bl.11.10, Bl.11.17 IM.2.12	Bl.4.5/Bl.4.7 (General medicine) Bl.11.9, Bl.11.10 (Perform level competency) Bl.11.17 (General medicine, Pathology) IM.2.12 (General medicine)
13.1.	Estimation of Total Cholesterol	Bl.11.9	Bl.11.9 (Perform level competency)
13.2.	Estimation of Triglyceride	Bl.11.10	Bl.11.10 (Perform level competency)
13.3.	Estimation of HDL Cholesterol	Bl.11.9	Bl.11.9 (Perform level competency)
	Case Study: Dyslipidemia and Myocardial Infarction		Early clinical exposure
14.	Estimation of Serum Calcium and Phosphorus	Bl.11.11	Bl.11.11 (Perform level competency)

S.No.	Experiment	Competency Covered	Early Clinical Exposure/ATCOM/ Integration (Vertical/Horizontal)/Certification
UNIT 4	**Self-directed Learning Exercises**		
15.	pH Meter and Measurement of pH of Buffers	BI.11.16	
16.	Water Electrolytes Homeostasis and Estimation of Na$^+$ and K$^+$ with ISE Analyzer	BI.11.16	
17.	Blood Gas Analysis by ABG Analyzer	BI.11.16	
18.	Paper Chromatography and Thin layer Chromatography (TLC)	BI.11.16	
19.	Protein Electrophoresis (PAGE)	BI.11.16	
20.	ELISA	BI.11.16	
21.	Immunodiffusion and Immunochemical Analysis	BI.11.16	
22.	Quality Control in Clinical Laboratory	BI.11.16	
	Hospital Visit to Central Diagnostic Laboratory		Early clinical exposure and reflective writing
23.	DNA Isolation from Blood and Tissue	BI.11.16	
Unit 5	**Early Clinical Exposure Exercises and Reflection Writing**		
24.	CSF Examination	BI.11.15	
	Hospital Visit to Pediatric Department		ECE and reflective writing
25.	Thyroid Function Test	BI.6.13, BI.6.14	BI.6.13, BI.6.14 (Physiology, anatomy, pathology, general medicine)
	Visit to Laboratory and Interview with Patients with Thyroid Disorders		ECE and reflective writing
26.	Pancreatic Function Test	PY.4.8, PY.8.4	PY.4.8, PY.8.4 (Physiology)
	Hospital Visit to Gastroenterology/Medicine Department		ECE and reflective writing
27.	Disorders of Acid-Base Balance	BI.11.17	BI.11.17 (Pathology, general medicine)
	Hospital Visit to Emergency/ICU Department		ECE and reflective writing
Unit 6	**Attitude, Ethics and Communication (AETCOM) Modules**		
28.	Introduction of Clinical Methods	BI.11.2	Integration of all three department of 1st professional year
29.	What does it Mean to be a Doctor?	AETCOM module 1.1	Integration of all three department of 1st professional year
30.	What does it Mean to be a Patient?	AETCOM module 1.2	Integration of all three department of 1st professional year
31.	Doctor-Patient relationship	AETCOM module 1.3	Integration of all three department of 1st professional year
32.	Foundation of Communication -1	AETCOM module 1.4	Integration of all three department of 1st professional year
Unit 7	**Biochemical Calculations and Reference Range**		
Unit 8	**Practical Spots in Biochemistry**		
Unit 9	**Competency-based Assessment**		

UNIT 1

Introduction of Clinical Biochemistry

OUTLINE

1.1. Laboratory Apparatus and Equipment, Good and Safe Laboratory Practice, and Waste Disposal Systems in Laboratory

Laboratory Apparatus and Equipment, Good and Safe Laboratory Practice, and Waste Disposal Systems in Laboratory

1.1

LEARNING OBJECTIVES

At the end of the session the student must be able to know about:
- Function and principle of commonly used laboratory apparatus as well as their clinical applications
- Good safe laboratory practice
- Laboratory waste disposal systems
- First aid procedure during laboratory work.

Competency BI.11.1: Describe commonly used laboratory apparatus, good safe laboratory practice, and waste disposal.
Competency BI.11.19: Outline the basic principles involved in functioning of instruments commonly used in a biochemistry laboratory and their applications.
Domain: Knows.
Level: Knows how.
Core competency: Yes.

DEVELOPMENT OF CLINICAL BIOCHEMISTRY LABORATORY

Clinical biochemistry is the branch of medical science, which deals with detection of biomarkers in body fluids for detection and management of various disease conditions.

Metabolites, solutes, enzymes, salts, acids, bases, and gases can be detected using appropriate method and technique for the diagnosis of metabolic disorders, genetic disorders, systemic diseases, and enzyme defects.

Clinical chemistry is the application of chemical techniques to study the diseases in humans. It includes investigations required for the purpose of:
- Diagnosis of disease
- Prognosis and management
- Screening test.

The function of clinical chemistry laboratory is to perform qualitative and quantitative analysis on body fluids such as blood, urine, spinal fluids, feces, secretions, tissues, calculi, etc.

Since the results are to be used by clinicians in diagnosis and management of disease, the tests must be performed as accurately as possible. This requires the use of sound analytical methods and good instrumentation.

COMMON EQUIPMENT USED IN CLINICAL CHEMISTRY LABORATORY

Since the development of clinical biochemistry, various instruments were developed for early and accurate diagnosis. In this chapter, we deal with development of various instruments in the history of biochemistry and their contribution in clinical biochemistry laboratory.

- **Balances:** An accurately operating balance is essential in producing high-quality reagents and standards.
  A balance may have one or two pans.
  a. **Double pan balances** consisting of a single beam with arms of equal length. Standard weights are usually added by hand to the right side pan to counterbalance the weight of the object on the other. It is used for weighing chemicals more than 1 g **(Fig. 1.1.1)**.
  b. **Single pan balances** have arms of unequal length. Object to be weighed is placed on the pan attached to the shorter arm. Restoring force is applied mechanically or electronically to the other arm to return the beam to its null position.
  c. **Electronic balances** are single pan balances that use electromagnetic force to return the balance beam

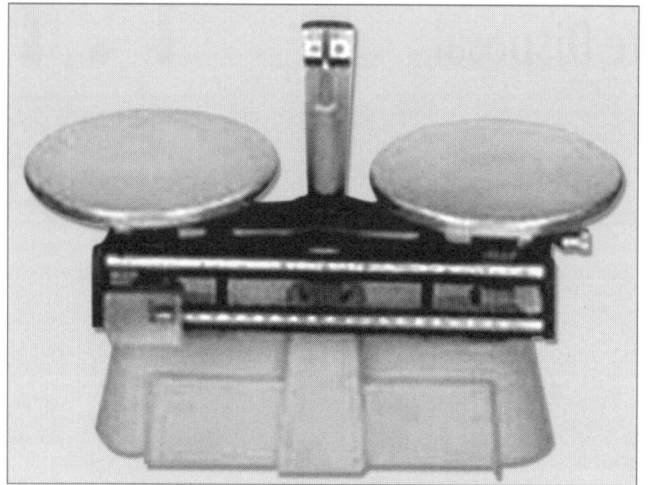

Fig. 1.1.1: Double pan balance.

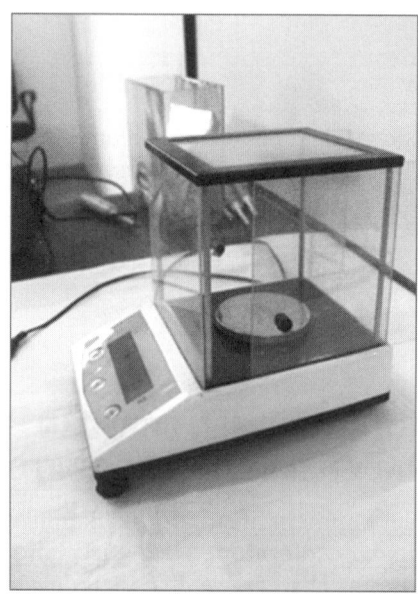

Fig. 1.1.2: Electronic balance.

Fig. 1.1.3: Hot air oven.

Fig. 1.1.4: Water bath.

to its null position. Almost all electronic balances have a built in provision for taring, so that the mass of the container can be subtracted easily from the total mass measured. These electronic balances are more sensitive and can be used to measure even microgram quantities. Balances should be kept scrupulously clean and kept in an area away from dust, vibrations, and other electrical equipment. It should not be shifted very often **(Fig. 1.1.2)**.

- ❖ **Hot air ovens:** These are used in laboratory for drying and sterilization of the glassware. Temperature range varies from 20°C to 250°C **(Fig. 1.1.3)**.
- ❖ **Water baths:** A temperature controlled device filled with deionized water used to warm and maintain materials at a specified temperature **(Fig. 1.1.4)**.

Use of deionized water with a bactericidal agent is necessary to prevent salt deposition on heat exchangers and also bacterial proliferation in water; such deposits interfere with maintenance of adequate temperature control.
- ➤ **Serological:** Temperature of water bath is maintained between 25°C and 40°C (mainly at 37°C).
- ➤ **Boiling:** Temperature of water bath is maintained at 100°C.
- ❖ **Centrifuges:** Centrifugation is a process whereby centrifugal force is used to separate solid matter from a liquid suspension. The centrifuge is an apparatus that carries out this action **(Fig. 1.1.5)**.

The speed is expressed in *revolutions per minute* (rpm) and the centrifugal force generated is expressed in terms of *relative centrifugal force (*RCF) or gravities.

Fig. 1.1.5: Centrifuge.

Classification of centrifuges may be based on several criteria, which may include: (a) Table-top or floor model, (b) Refrigeration, (c) Type of head (fixed, hematocrit, or angled), and (d) Maximum speed attainable (e.g. ultracentrifuge).

- **Pipettes:** Calibrated glass pipettes are available in various volumes ranging from 0.1 mL to 20 mL. Glass pipettes may be **delivery/blow-out** and **non-blow out.** Blow-out pipettes have volume graduations that extend to the delivery tip of the pipette. Thus, last blown out drop of liquid is also included in the delivery volume. Automatic pipetting devices have nowadays replaced the manual glass or plastic pipettes. The major advantages of the automatic pipettes are accuracies, time-saving, safety and precision, ease of use, stability, and lack of required cleaning, since the contaminated portions of the pipette such as the tips are often disposable **(Fig. 1.1.6)**.
- **Plasticware:** Nowadays, most of glass pipettes replaced by automated plasticware made up of high-density polyethylene and teflon.

  The automatic pipettes may be:
  - **Micropipettes**—volume 1–1,000 µL (1 mL)
  - **Macropipettes**—volume > 1 mL
  - They are available either in fixed volume or with variable volume adjustment.

  Note: Use only volume-specific pipettes. Try to use the pipettes whose volume is near to the volume to be used in that experiment, e.g. to take 1 mL, use 1 mL pipettes; if that is not available, use nearest volume that is 2 mL. Try to avoid using 5 mL or 10 mL pipettes for 1 mL volume.

- **Colorimeters and spectrophotometers:** Most commonly used photometric instruments in present day's clinical chemistry laboratory **(Figs. 1.1.7 and 1.1.8)**. They are indispensable for a clinical biochemistry laboratory.
- **Automated analyzers:** Improved healthcare facilities and high workload on clinical laboratory that is why, there is need of high throughput automated analyzers with high accuracy and precision of laboratory results.
- **Semiautomated analyzers:** Improved version of colorimeter and spectrophotometers with significant degree of work simplification with digital readout and printing devices.
- **Fully automatic analyzers:** Computer and robotic techniques are used for highly efficient and high throughput results, ideally suitable for large sample workload **(Fig. 1.1.9)**.

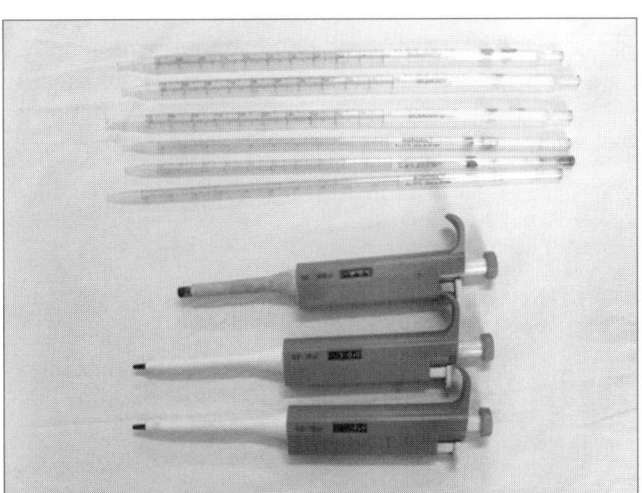

Fig. 1.1.6: Pipettes.

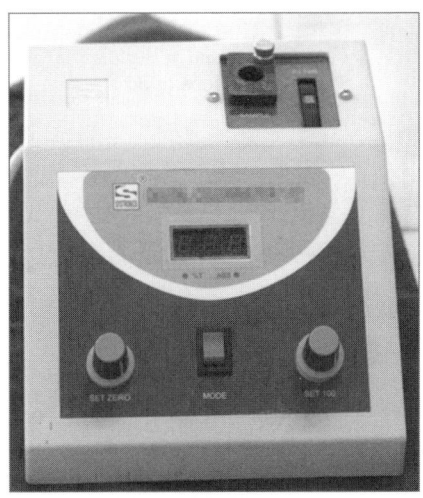

Fig. 1.1.7: Colorimeter.

Unit 1: Introduction of Clinical Biochemistry

Fig. 1.1.8: Spectrophotometer.

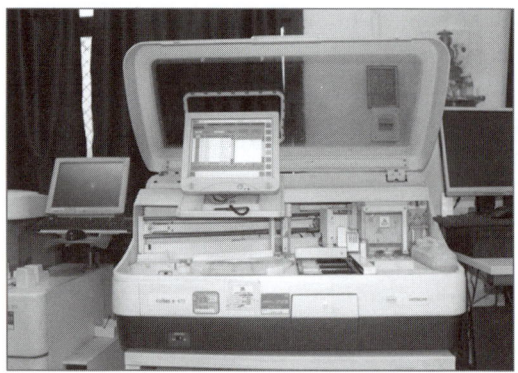

Fig. 1.1.9: Fully automatic analyzers.

SAMPLE COLLECTION AND HANDLING

A. Different Vials used in Clinical Diagnostic Laboratory

Tube	Cap color	Additive	Determination	Inversion
Sodium citrate	Light blue	Buffered sodium citrate 0.109M (3.2%)	PT, APTT, D-Dimer, Lupus Anticoagulant, Anti-Thrombin III, Protein S and C, Fibrinogen, Factor Assay, HLA B27	3–4
Red top plain	Red	Clot activator silicone-coated	Cold Agglutinins, Abnormal blood group Antibody Screen, Antibody Identification and Vitamin D2 and D3 (HPLC)	5
Plain	Gold	Clot activator and gel for serum separation	Biochemistry, Virology, Immunology, Vit B12, Folate, Anticonvulsant Drug Monitoring, Therapeutic Drug Monitoring (TDM) and other investigations that require serum (i.e. plain blood)	5
Royal blue plain	Royal blue (Serum)	Trace element (With clot activator)	Zinc, Copper, Aluminum, Selenium	8–10
Lithium heparin	Green	Lithium heparin	STAT Biochemistry, Plasma Ammonia (in ice pack), CD34, Blood Alcohol, Toxicological Tests, Chromosomes Studies, Electrolytes, Renal Function Tests, TB Spot	8–10
Lithium heparin	Light green	Lithium heparin and gel for plasma separation	STAT Biochemistry, Plasma Ammonia (in ice pack), CD34, Blood Alcohol, Toxicological Tests, Chromosomes Studies, Electrolytes, Renal Function Tests	8–10
Royal blue EDTA	Royal blue (K_2 EDTA)	K_2 EDTA	Arsenic, Cadmium, Chromium, Mercury, Lead, Manganese	8–10
K_2 EDTA	Lavender	Spray-coated K_2 EDTA	FBC, PBF, MP, ESR, HBA1c, cyclosporine A, G6PD, RBC cholinesterase, ACTH, Renin, HB electrophoresis, thalassemia screening, CD4/CD18, factor V leiden	8–10
Sodium fluoride	Gray	Sodium fluoride	Glucose, lactic acid (in ice pack)	8–10

Courtesy: Western Diagnostic Laboratory, Myaree, Australia.

B. Rejection and Acceptance Criteria of a Blood Sample

Acceptance	Rejection
Complete request form	Improper test request • Incomplete, duplicate • Inconsistent information • Error in test input
Patient identification and sample labeled properly	Misidentification • Unlabed sample • Wrongly labeled sample • Mismatch sample
Suitable vial selected	Improper container or vial
Sufficient amount of blood	Insufficient sample • Dried sample • Leaking vial • Insufficient quantity of sample
Sample stored and transported properly	Inappropriate transport • Temperature during transport • Light exposure • Prolonged /delayed transport • Incorrect preservation and storage of sample
Sample according to specific test requisition (Fasting sample, pooled sample, plasma in anticoagulant vial)	Inappropriate sample • Inappropriate concentration • Inappropriate composition • Bacterial contamination • Hemolyzed sample • Lipemic sample • Clotted sample with fibrin (specially in hematological test and coagulation profile)

GOOD AND SAFE LABORATORY PRACTICE

Cardinal Rules to be followed in Laboratory

- ❖ **Good personal behavior/habits:**
  - ➢ Students must wear apron whenever they are in the laboratory for practical class.
  - ➢ Students should bring glass marking pencil, notebook, and practical journal during practical classes.
  - ➢ Maintain silence in the laboratory.
  - ➢ Long hair should not be left loose.
  - ➢ Do not eat or drink or smoke in work area.
- ❖ **Good housekeeping:**
  - ➢ Students should keep working table clean during and after their work, i.e. work area should be kept free of chemicals, dirty, and broken glassware.
  - ➢ Replace the chemicals, glassware, and reagent bottles at the appropriate places.
- ❖ **Good laboratory technique:**
  - ➢ Students should read the relevant data regarding the experiments to be performed.
  - ➢ All instructions and labels in the laboratory should be carefully read.
  - ➢ New or unfamiliar equipment should not be operated until you have been given instructions about it.
  - ➢ Strong acids/alkalis should never be mouth pipetted. Autopipettes and dispensers should be used. Make sure that acid/alkali should not fall on your hands or on any part of the body.
  - ➢ Be careful when you boil the solutions.
  - ➢ Perform and preserve all the tests in serial order duly labeled.
  - ➢ Note down your observations of:
    - ♦ Qualitative experiments as test, observation, and inference.
    - ♦ Quantitative experiments as aim, principle, observation, calculation, reference range, and clinical significance.
  - ➢ Note down your observation and get them verified by the teacher.
  - ➢ Report breakages to the technicians/teachers.

Laboratory Hazards

- ❖ **Biological hazards:** Every patient's specimen should be treated as potentially infectious. Blood samples from high-risk patients (like AIDS, hepatitis B, etc.) should be collected, transported, handled, and processed using strict precautions. Gloves, masks, and gowns should be worn. Specimens should remain "capped" during centrifugation to prevent formation of infective aerosols.
- ❖ **Physical and chemical hazards:** Careless handling of apparatus and reagents is a common cause of laboratory accidents resulting in burns or fire that must be reported and treated promptly as per the chart displayed in the laboratory. In clinical chemistry laboratory, all elements essential for fire to begin are present: fuel, heat, or ignition source (chemicals and electricity) hence, it is very important to assure safety of self, personnel, and equipment.

DISPOSAL OF LABORATORY WASTE

Besides the hazard of burns/fires, workers in clinical laboratory are exposed to biological hazards due to handling of infected specimen. Hence, proper safety is required not only in handling, but also in disposal of these infected materials.

It has been recommended that waste should be segregated at the point of generation and disposed in bags with correct color coding.

Black bags (for municipal dumps)—paper, peels, and wrappers (noninfective material).

Yellow bags (for incinerator)—swabs and items contaminated with blood and body fluids, discarded medicines, gloves, etc.

Blue bags (for shredder)—syringes, needles, and sharps are first destroyed in needle destroyer and discarded in sharp disposal unit containing 1% bleach and then segregated in cardboard boxes (puncture-proof).

FIRST AID

- Inhalational injury by toxic fumes is best treated by removal to an uncontaminated atmosphere and treating for shock. Irritation of throat can be relieved by warm soothing drinks or by hot water vapor inhalation.
- Chemical injury to eyes should be washed for several minutes using a gentle stream of water.
- Skin burns should be washed under running water and then petroleum jelly, olive oil, or burn ointment must be applied and covered with dry sterile gauze.
- In accidental swallowing, mouth should be repeatedly rinsed with water. If proper swallowing has occurred, the person should be given water to drink followed by milk in case of acids and dilute lime juice in case of alkalis. ***Sodium bicarbonate should not be given***.
- Wounds with contamination of infected materials are rinsed with water and washed with soap solution before applying antiseptic solutions.
- Injuries by broken glassware are treated by washing and removing the glass piece. Bleeding should be stopped by compression of nearby body parts with tight bandage.

Mercurochrome/Acriflavine ointment may be applied and it has to be covered with gauze.

VIVA VOCE QUESTIONS

1. Universal precautions for health care workers.
2. Enlist different anticoagulants used during the collection of blood for various investigations.
3. Classify different type of centrifuge used in clinical biochemistry laboratory.
4. What are the acceptance and rejection criteria for blood sample at clinical biochemistry laboratory?

PRACTICAL ASSIGNMENT

1. Separate serum from blood sample with the help of centrifuge.
2. Weighing of salt powder (different weight range) with the help of electronic balance.
3. Calibration of pipettes used in clinical laboratory.

NOTES

UNIT 2

Qualitative Experiments and their Clinical Applications

OUTLINES

2.1. Analysis of Carbohydrates
2.2. Analysis of Proteins
2.3. Analysis of Physical and Chemical Composition of Physiological Urine
2.4. Identify, Perform and Interpret Pathological Urine Analysis and Correlate it with Pathological States

Analysis of Carbohydrates

2.1

LEARNING OBJECTIVES

At the end of the session the student must be able to:
- Differentiate between different types of carbohydrate important in health and diseases
- Detect different types of carbohydrate by qualitative analysis
- Detect unknown carbohydrate present in solutions
- Interpret different quantitative laboratory investigations related to disorders of carbohydrate metabolism.

Competency BI.3.1: Discuss and differentiate monosaccharides and polysaccharides giving example of main carbohydrates as energy fuel, structural element, and storage in the human body.
Competency BI.3.8: Discuss and interpret results of analytes associated with metabolism of carbohydrates.
Domain: Knows.
Level: Knows how.
Core competency: Yes.

INTRODUCTION

Carbohydrates are defined as polyhydroxy aldehydes or ketones or their derivatives. They are classified into monosaccharides, disaccharides, oligosaccharides, and polysaccharides.

The following carbohydrates are generally given for practical exercises: glucose, fructose, maltose, lactose, sucrose, starch, and dextrin.

CLASSIFICATION

Broadly classified as:
- **Monosaccharides:** Simplest sugars, which cannot be hydrolyzed further into smaller compounds, e.g. glucose, fructose.
- **Disaccharides:** They are formed by condensation of two monosaccharide molecules, which may be similar or dissimilar, e.g. maltose, lactose, and sucrose.
- **Polysaccharides:** They are formed by condensation of large numbers of monosaccharide molecules. On hydrolysis, they give more than ten monosaccharide units, which may be similar or dissimilar, e.g. starch, glycogen, heparin, hyaluronic acid, etc.

MOLISCH'S TEST

This is the group test for carbohydrates given positive by all carbohydrates whether free or bound to such substances as protein (glycoprotein and mucoprotein) and lipids (glycolipids) **(Fig. 2.1.1)**.

Experiment	Observation	Inference
Molisch's test: 2 mL of original solution + 2 drops of 1% alcoholic α-naphthol (Molisch's reagent). Mix and add 2 mL of concentrated H_2SO_4 slowly from the side of the test tube	Purple ring at the junction of the two liquids	*Carbohydrates present:* It is a general test, which indicates the presence of carbohydrates

Principle

Carbohydrates undergo dehydration when treated with concentrated H_2SO_4 to form a five-membered cyclic

Fig. 2.1.1: Molisch's test.

compound, furfural or its derivative, which on condensation with α-naphthol forms a purple-colored complex.

Fallacies

Furfural, aldehydes, and some organic acids, e.g. formic acid, oxalic acid, lactic acid, citric acid, etc. also give a positive Molisch's test **(Fig. 2.1.1)**.

Precautions

* Test tube must be dry.
* Do not shake the test tube to get a positive test.

IODINE TEST

This is the test for polysaccharides. Starch, dextrin, and glycogen can be differentiated by this test.

Experiment	Observation		Inference
3 mL of original solution + 2 drops of iodine solution Mix.	No color develops		Polysaccharides absent
	Color	On heating / On cooling	
	Blue	Disappears / Reappears	Starch
	Reddish purple	Disappears / Reappears	Dextrin
	Reddish brown	Disappears / Reappears	Glycogen

Principle

The test depends upon the property of adsorption possessed by the large polysaccharide molecules, which adsorb the smaller iodine molecules on their surface to form a complex of undefined chemical nature. The starch-iodine, dextrin-iodine, and glycogen-iodine complexes are blue, reddish purple, and reddish brown, respectively. The adsorption is a physical property, which decreases on heating, the complex dissociates and color disappears. On cooling, the complex is reformed, and color reappears, except in the case of dextrin in which heating produces a nearly irreversible change, as a result it loses the property of adsorption. Monosaccharides and disaccharides do not produce a change in color **(Fig. 2.1.2)**.

Precautions

* The test should not be carried out in alkaline medium as iodine is not freely available and iodine adsorption does not occur.
* Excess of iodine should be avoided.
* To prevent hydrolysis, the solution must not be heated highly.

TEST FOR THE REDUCING SUGARS

The carbohydrates, which have a free aldehyde and ketonic groups, have the ability to reduce the solutions of various metallic ions.

Requirement

* *Reducing agent:* Carbohydrate having free aldehyde and ketone groups.
* Oxidizing agents having metallic ions, e.g. Cu^+, Cu^{2+}, Ag^+, etc. can be used as salt.
* *Medium:* Alkaline medium for Fehling's and Benedict's test. Acidic medium for Barfoed's test.
* Heating agent.

Benedict's Test

This is a test for the reducing carbohydrates **(Fig. 2.1.3)**.

Test	Observations	Inference
Benedict's test: 5 mL of Benedict's reagent + 8 drops of original solution. Boil it for 2 minutes and allow it to cool	The color of the precipitate could be green/yellow/orange/red/brick red depending on the concentration of reducing sugar in the solution	–

Principle

In the alkaline medium, cupric ions of the Benedict's solution are reduced to cuprous ions by the reducing sugars, which ultimately form Cu_2O as a red precipitate.

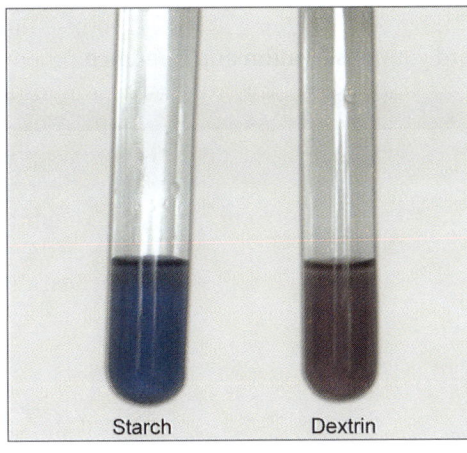

Fig. 2.1.2: Iodine test.

Fig. 2.1.3: Benedict's test.

Advantages

- The reagent is very stable and can be stored in a single bottle for a long time.
- The test is convenient and simple.
- Since, sodium carbonate is a very weak base, the carbohydrates are not destroyed.
- This is **semi-quantitative** test. If the solutions are taken in proper proportion and procedure is followed strictly, the approximate concentration can be judged from the color of precipitation.

Grades	Color of precipitation	Approximate concentration of carbohydrates
0	Blue	No sugar
(+)	Green	0.1–0.5%
(++)	Yellow	0.5–1.0%
(+++)	Orange	1.0–2.0%
(++++)	Red	Above 2.0%

Practical Application

This is the most widely employed test for the detection of reducing sugar in urine.
Lactosuria — Lactose
Glucosuria — Glucose
Galactosuria — Galactose
Pentosuria — Pentose
Fructosuria — Fructose

Disadvantages

- Many substances with reducing groups react with cupric ion in alkaline medium at boiling temperature to form red precipitates of cuprous oxide.
  Noncarbohydrates giving positive Benedict's test:
  - High concentration of uric acid, creatinine, and ketones
  - Homogentisic acid (solution turns black due to black-colored oxidized homogentisic acid)
  - Vitamin C (even without boiling)
  - Certain drugs like aspirin, cephalosporins.
- Reagent is unstable. It has to be prepared in 2 parts, which are stored separately.
- The strong alkali as KOH is present in the reagent, which can destroy carbohydrate.
- Autoreduction may occur resulting in the false-positive test.

Unit 2: Qualitative Experiments and their Clinical Applications

Exercise 2.1.1: Analysis of carbohydrates.

Perform the following tests on sugar solution provided to you and record your observation and inference.

Test	Observation	Inference

Result and Interpretation

....
....
....
....
....
....
....

Date: Teacher's signature

Analysis of carbohydrates.

Perform the following tests on sugar solution provided to you and record your observation and inference.

Test	Observation	Inference

Result and Interpretation

....
....
....
....
....
....
....

Date: **Teacher's signature**

Barfoed's Test

This test is for strong reducing monosaccharide **(Fig. 2.1.4)**.

Test	Observation	Inference
2 mL of Barfoed's reagent + 2 mL of original solution. Boil it for 30 seconds and allow it to cool	Scanty red precipitate sticking to the side and bottom of the test tube	Monosaccharide present. This test differentiates between monosaccharides and disaccharides. Disaccharides give this test negative

Principle

The reduction of cupric ions is carried out in acidic medium. Since acidic medium is unfavorable for reduction, only the strong reducing carbohydrates, i.e. monosaccharide give the test positive.

Precautions

1. Boiling period should not exceed 2 minutes otherwise disaccharide will hydrolyze and give positive test.
2. Chloride ions interfere with this test therefore, the test should not be carried out with solution containing chloride, i.e. urine.

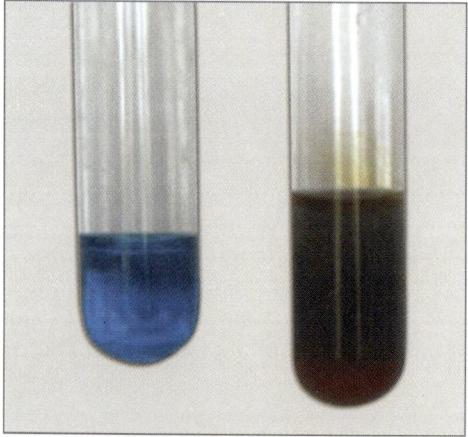

Fig. 2.1.4: Barfoed's test.

Seliwanoff's Test

This test is for ketohexoses, i.e. fructose **(Fig. 2.1.5)**.

Test	Observation	Inference
5 mL of Seliwanoff's reagent + 4–5 drops of original solution. Boil for 30 seconds and allow it to cool	Cherry-red color	Ketose sugar present. The test distinguishes between aldose and ketose sugars

Principle

The carbohydrates are converted into furfural derivatives by HCl, which condense with resorcinol to form cherry red or pink color complex.

Precautions

- ❖ Boiling should not exceed more than 60 seconds as prolonged boiling converts aldohexose to ketohexose giving false-positive test.
- ❖ The final concentration of HCl in Seliwanoff's reagent should not exceed 12% as strong acids convert aldohexoses to ketohexoses giving a positive test.

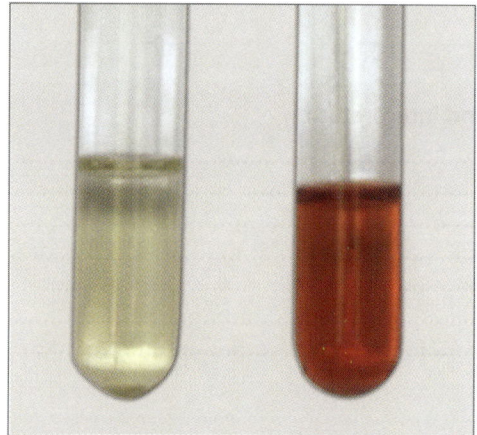

Fig. 2.1.5: Seliwanoff's test.

Exercise 2.1.2: Analysis of carbohydrates.

Perform the following tests on sugar solution provided to you and record your observation and inference.

Test	Observation	Inference

Result and Interpretation

....
....
....
....
....
....
....

Date: Teacher's signature

Unit 2: Qualitative Experiments and their Clinical Applications

Analysis of carbohydrates.

Perform the following tests on sugar solution provided to you and record your observation and inference.

Test	Observation	Inference

Result and Interpretation

....
....
....
....
....
....
....

Date: **Teacher's signature**

Osazone Test

This is a test for the identification of reducing disaccharide and many monosaccharides by characteristic crystal formation. It is a confirmatory test **(Fig. 2.1.6)**.

Test	Observation	Inference	
Take 5 mL of glucose, fructose, maltose, and lactose solution in labeled test tube. Add 0.3 g of Osazone mixture + 5 drops of acetic acid. Put all test tubes in water bath	Yellow crystals will appear in the tubes containing glucose and fructose within 5 minutes and maltose and lactose after 45 minutes. Put crystal on slide and observe under microscope	Glucose	Needle shape
		Fructose	Needle shape
		Maltose	Sunflower
		Lactose	Puff shaped

Principle

When the reducing carbohydrates are treated with phenylhydrazine at 100°C and pH of 4.3, formation of osazones of their respective carbohydrates takes place with difference in: (A) Time of formation, (B) Solubility in boiling water, (C) Melting point, and (D) Crystalline structure.

Inversion Test

This is a test for nonreducing sugar sucrose.

Test	Observation	Inference
Inversion test for sucrose: 3 mL of original solution + 2–3 drops of concentrated HCl. Boil for 2 minutes. Cool, neutralize with 2–3 drops of 40% NaOH or sodium carbonate till effervescence ceases. Then, perform: **i. Benedict's test and** **ii. Seliwanoff's test**	Precipitate of green/yellow/orange/brick red Cheery-red color	Sucrose when boiled with concentrated HCl undergoes hydrolysis to yield glucose and fructose, which being reducing sugars give Benedict's test positive. Ketohexose present

Principle

When sucrose is boiled with concentrated HCl, it is hydrolyzed to form monosaccharide units of glucose and fructose.

$$\text{Sucrose} \longrightarrow \text{Fructose} + \text{Glucose}$$
$$+66.5° \qquad\qquad -92.3° \qquad\qquad +52.5°$$

The optical rotation changes from dextrorotatory to levorotatory as fructose possess a much stronger levorotation than dextrorotation caused by glucose. The change in optical rotation is, thus, known as **inversion**. So, sucrose, a mixture of glucose and fructose, is also called as **invert sugar**.

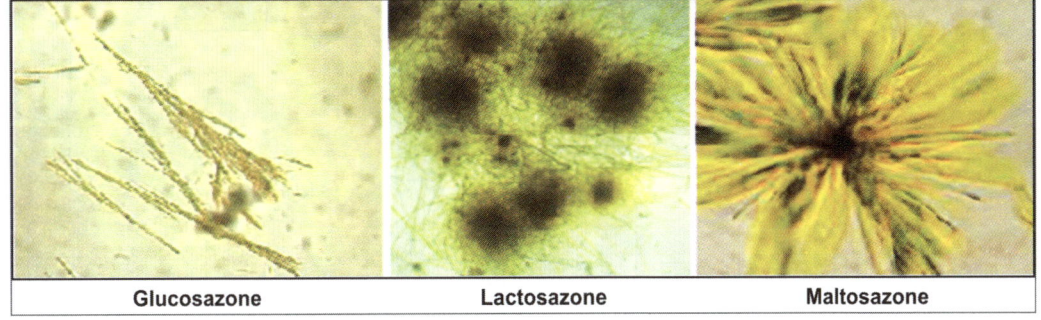

Fig. 2.1.6: Shape of osazones under microscope.
Courtesy: DM Vasudevan. Textbook of Biochemistry for Medical Students. 8th edn. Jaypee Brothers Medical Publishers (P) Ltd, New Delhi, India.

Unit 2: Qualitative Experiments and their Clinical Applications

SCHEME FOR IDENTIFICATION OF UNKNOWN CARBOHYDRATE

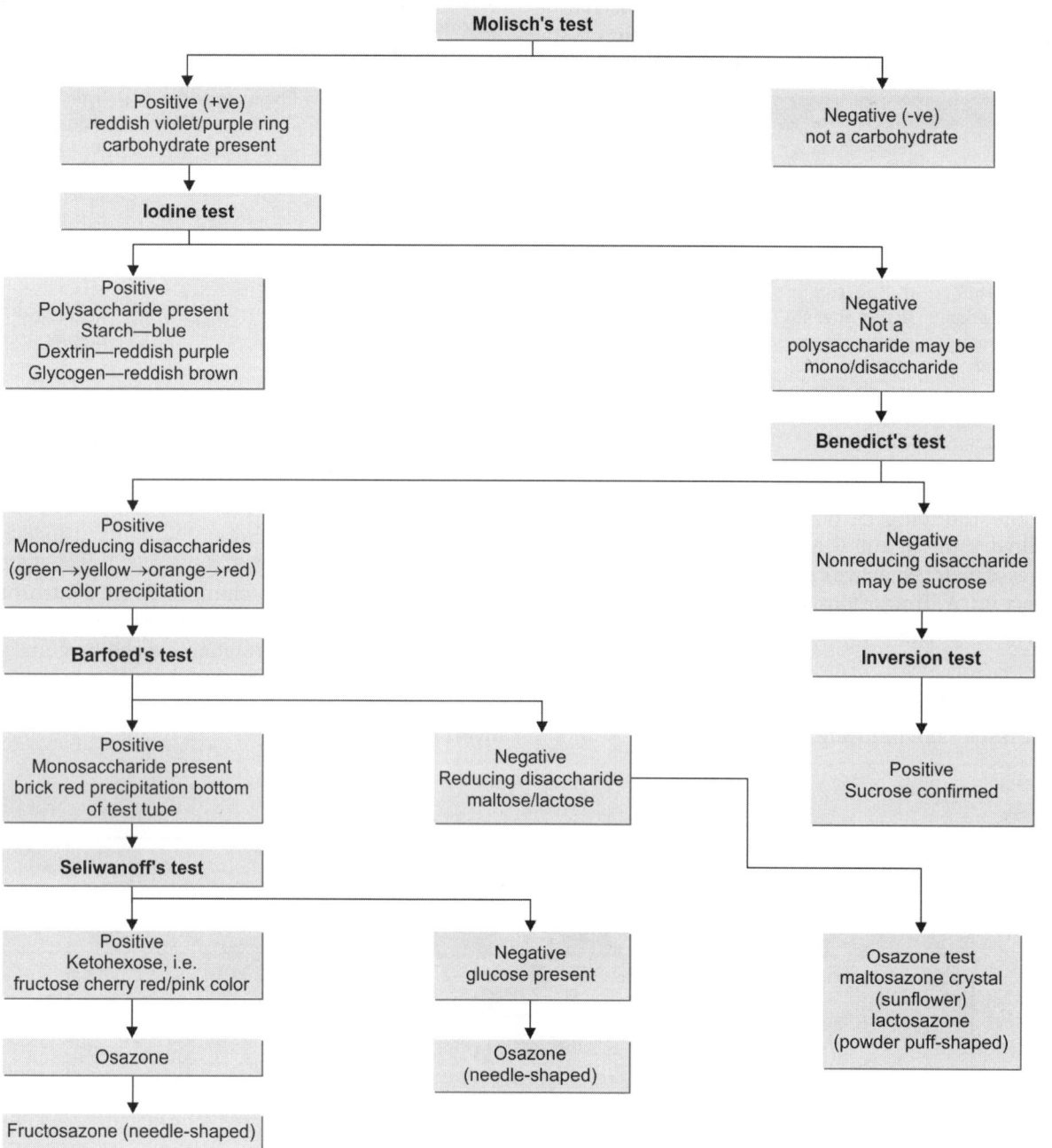

Exercise 2.1.3: Analysis of carbohydrates.

Perform the following tests on sugar solution provided to you and record your observation and inference.

Test	Observation	Inference

Result and Interpretation

....
....
....
....
....
....
....

Date: **Teacher's signature**

Unit 2: Qualitative Experiments and their Clinical Applications

Analysis of carbohydrates.

Perform the following tests on sugar solution provided to you and record your observation and inference.

Test	Observation	Inference

Result and Interpretation

....
....
....
....
....
....
....

Date: **Teacher's signature**

VIVA VOCE QUESTIONS: CARBOHYDRATES

1. Whether all carbohydrates give Molisch's test positive. Will glyceraldehydes 3-carbon aldose give Molisch's test positive?
2. Benedict's test is a semi-quantitative test. Why?
3. Why glucose, fructose, give same type of crystals during osazone formation?
4. Name the typical carbohydrates found in the plants and animals.
5. Starch gives blue color and glycogen gives red color with iodine test. Why? Why on heating color disappears in starch-iodine solution and reappears on cooling?

NOTES

Analysis of Proteins

2.2

> **LEARNING OBJECTIVES**
>
> At the end of the session the student must be able to:
> - Differentiate between different types of protein important in health and diseases
> - Detect different types of protein by qualitative analysis
> - Detect unknown protein present in solutions
> - Interpret different quantitative laboratory investigations related to disorders of protein metabolism.

Competency BI.5.4: Describe common disorders associated with protein metabolism.
Competency BI.5.5: Interpret laboratory results of analytes associated with metabolism of proteins.
Domain: Knows.
Level: Knows how.
Core competency: Yes.

INTRODUCTION

Proteins are made up of a large number of different α-amino acids joined together by peptide linkages. All proteins on hydrolysis with strong acid or enzymes yield amino acids.

Presence of different functional groups in various amino acids gives different color reaction of protein.

CLASSIFICATION

Broadly classified as:
Simple proteins: Contains only amino acids, e.g. albumin, globulin, protamines, and prolamins.

Conjugated proteins: Combination of protein with a nonprotein part called prosthetic group, e.g. glycoproteins, lipoproteins, nucleoproteins, chromoproteins, and metalloproteins.

Derived proteins: Degradation products of native protein, e.g. peptones, peptides, and amino acids.

Identification of Unknown Protein

Biuret test: This test is given by all peptides and proteins having at least two peptide bonds, hence it is used as a general test for peptide and proteins.

Test	Observation	Inference
Biuret test: 3 mL of original solution + 2 mL of 5% NaOH + 2 drops of 1% $CuSO_4$. Mix well. **Control:** 3 mL of distilled water + 2 mL of 5% NaOH + 2 drops of $CuSO_4$. Mix well.	Violet color Blue color	Two or more **peptide linkages are present. Principle:** In an alkaline medium, peptide linkage present in protein reacts with dilute copper sulfate to form a violet-colored complex. This is due to the coordination of cupric ions with the unshared electron pairs of peptide nitrogen and the oxygen. **Since two or more peptide linkages are required for the test, dipeptides do not give the test positive.** When urea is heated to about 180°C, it decomposes to form biuret, which gives this test positive; hence, the name

Heat coagulation test: When protein is heated, weak bonds like hydrogen, van der Waals, etc. are broken and protein is said to be denatured **(Fig. 2.2.1)**.

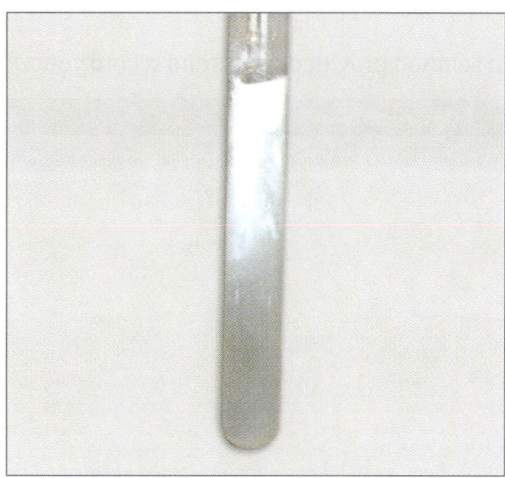

Fig. 2.2.1: Heat coagulation test.

Test	Observation	Inference
Heat coagulation test: Fill-up three-fourths of test tube with original solution. Heat the upper part of the solution. Add 2–3 drops of 1% acetic acid.	White precipitates are formed (protein coagulum is not dissolved in acetic acid).	**Principle:** When a protein is heated, its physical, chemical, and biological properties are changed due to breaking up of certain bond and the resultant change in the conformation of its molecules. This process is known as denaturation. However, when the coagulable proteins are heated at their isoelectric pH, a series of changes occur involving dissociation of the protein subunits (disruption of quaternary structure), uncoiling of the polypeptide chains (disruption of tertiary and secondary structure), and matting together of the uncoiled polypeptide chain (coagulation). While a denatured protein may be restored to its original structure and function by certain manipulations, coagulation is an irreversible process.

Unit 2: Qualitative Experiments and their Clinical Applications

Exercise 2.2.1: Analysis of protein.

Perform the following tests on protein solution provided to you and record your observation and inference.

Test	Observation	Inference

Result and Interpretation

....
....
....
....
....
....
....

Date: **Teacher's signature**

Analysis of protein.

Perform the following tests on protein solution provided to you and record your observation and inference.

Test	Observation	Inference

Result and Interpretation

....
....
....
....
....
....
....

Date: **Teacher's signature**

Saturation and Precipitation Reactions of Proteins

Saturation Tests

Principle: Proteins, which are colloidal in nature, are kept in solution by two factors:
i. **Electric charges:** A large number of electric charges are present on the surface of protein molecules. The similarly charged particles repel each other and this prevents their coalescence.
ii. **Shell of hydration:** Each molecule is surrounded by a film of water known as the shell of hydration. The shell of hydration also prevents coalescence of particles. If both these factors are removed, the particles coalesce, and are precipitated. Adding a neutral salt such as ammonium sulfate, which neutralizes the electric charges, and removes the shell of hydration as it has got greater affinity for water than the colloid. The amount of ammonium sulfate required to precipitate a colloid depends upon the surface area of the particles. The larger the surface area, the larger the number of electric charges and the larger the shell of hydration.

Thus, small molecules, e.g. albumin, having a relatively large surface area are precipitated only by full saturation with ammonium sulfate. The larger molecules, e.g. casein and gelatin, have a smaller surface area and are, therefore, precipitated both by half saturation and full saturation with ammonium sulfate. Peptone, which has got very small molecules, is not precipitated even by full saturation with ammonium sulfate.

Test	Observation	Inference
Half saturation test: 5 mL of solution + 5 mL of saturated $(NH_4)_2SO_4$ solution. Mix and filter the precipitate. Biuret test with filtrate. **Biuret test:** 2 mL of filtrate + 2 mL of 40% NaOH + 2–3 drops of $CuSO_4$ solution. **Control:** 2 mL of distilled water + 2 mL of 40% NaOH + 2–3 drops of $CuSO_4$ solution.	**Positive:** Precipitates are formed and no violet color is seen in Biuret test. **Negative:** No precipitates are formed and violet color is seen in Biuret test.	Gelatin and globulin can be precipitated by half saturation. So, they give positive half saturation test. Albumin and peptone cannot be precipitated by half saturation. So, they give negative half saturation test.
Full saturation test: To 5 mL of solution, add $(NH_4)_2SO_4$ powder and shake to dissolve. Continue the process till it stops dissolving. Filter precipitation and then perform Biuret test with filtrate. **Biuret test:** 2 mL of filtrate + 2 mL of 40% NaOH + 2–3 drops of $CuSO_4$ **Control test:** 2 mL of DW + 2 mL of 40% NaOH + 2–3 drops of $CuSO_4$	**Positive:** White precipitation is formed and no violet color is seen in Biuret test. **Negative:** No precipitation is formed and/or violet color is observed in Biuret test.	Peptone is not precipitated by full saturation test. **Globulins are precipitated by half saturation while albumin is precipitated by full saturation of ammonium sulfate. The test is used for differential fractionation of albumin and globulin.** Addition of salts and electrolyte leads to adsorption of salvation envelope from the colloidal particle of protein solution along with neutralization of surface charges Albumin—smaller in size—larger surface area—hold more water molecule around it—required high salt concentration for saturation

Exercise 2.2.2: Analysis of protein.

Perform the following tests on protein solution provided to you and record your observation and inference.

Test	Observation	Inference

Result and Interpretation

....
....
....
....
....
....
....

Date: **Teacher's signature**

Unit 2: Qualitative Experiments and their Clinical Applications

Analysis of protein.

Perform the following tests on protein solution provided to you and record your observation and inference.

Test	Observation	Inference

Result and Interpretation

....
....
....
....
....
....
....

Date:　　　　　　　　　　　　　　　　　　　　　　　　　　　　　　　　　　**Teacher's signature**

Precipitation test: Any factor which removes protein charges or its shell of hydration, it can lead to precipitation of proteins.

Clinical Importance of Precipitation Reactions

- Protein-free filtrate is used for the estimation of blood sugar, urea, etc.
- Proteins are ppt. by salts of heavy metals of $HgCl_2$, $AgNO_3$, $CuSO_4$, etc.
- Are ppt. by certain acids, which are called as **alkaloidal reagents**, e.g. picric acid, phosphotungstic acid, and metaphosphoric acid.
- By concentrated solution of $(NH_4)_2SO_4$, Na_2SO_4, and Na_2CO_3 (salting-out method).
- Precipitation by dehydrating agents (alcohol, acetone). The agents convert them into suspensoids, which flocculate upon the addition of few drops of salt solution. Alcohol causes denaturation of proteins and it brings protein solution to isoelectric point at which it is precipitated.

Test	Observations	Inference
Precipitation by concentrated mineral acid: In a tube, take 3 mL of concentrated HNO_3 + 3 mL of original solution from the side of the tube	White ring at the junction of the two liquids	Concentrated mineral acids such as HNO_3, HCl, and H_2SO_4 denature the proteins and precipitate. **The test is used for the detection of albumin in urine**
Precipitation by heavy metal ions: 3 mL of original solution + 2 drops of 2% Na_2CO_3 solution + 1 drop of lead acetate solution	White precipitate	At pH above their pI, proteins are present as negatively-charged ions. The positively-charged metal ions neutralize the negative charge on the proteins and cause their precipitation. Commonly salts like lead acetate, ferric chloride, and cupric sulfate are used as precipitating reagents
Precipitation by alkaloidal reagents: i. 3 mL of original solution + 5 drops of 10% trichloroacetic acid ii. 3 mL of original solution + 5 drops of 20% sulfosalicylic acid	White precipitate	At pH below their pI, proteins are present as positively-charged ions. Alkaloids having negative charge neutralize the positive charge on the proteins and precipitate them. Trichloroacetic acid, Esbach's reagent (picric acid in citric acid), and sulfosalicylic acid are commonly used alkaloids. **Trichloroacetic acid test** is used for the preparation of protein-free filtrate of biological fluids. **Sulfosalicylic acid** test is used for the detection of albumin in the urine
Precipitation by alcohol: 3 mL of original solution + 1 mL of absolute alcohol	White precipitate	Organic solvents, e.g. ethanol (alcohol) and acetone reduce the water molecules available to proteins and also reduce the dielectric constant of the medium, precipitating proteins process

Unit 2: Qualitative Experiments and their Clinical Applications

Exercise 2.2.3: Analysis of protein.

Perform the following tests on protein solution provided to you and record your observation and inference.

Test	Observation	Inference

Result and Interpretation

....
....
....
....
....
....
....

Date: **Teacher's signature**

Analysis of protein.

Perform the following tests on protein solution provided to you and record your observation and inference.

Test	Observation	Inference

Result and Interpretation

....
....
....
....
....
....
....

Date: **Teacher's signature**

Color Reactions of Proteins (Fig. 2.2.2A to G)

Test	Observations	Inference
Ninhydrin test: 2 mL of original solution + 4–5 drops of 1% freshly prepared ninhydrin solution. Boil it.	Bluish-purple color	Free α-amino group is present. **Principle:** Ninhydrin reacts with free α-amino group present in amino acids to form hydrindantin and ammonia. The blue-colored complex—**Ruhemann's purple** is formed by the reaction of ninhydrin with hydrindantin and ammonia. Proline and hydroxyproline give yellow color as they do not have α-amino group.
Xanthoproteic test: 3 mL of original solution + 1 mL of concentrated HNO_3. Boil, cool, and add 40% NaOH.	The precipitate turn yellow on boiling, which turns orange in alkaline medium	Aromatic amino acid is present. **Principle:** The xanthoproteic reaction is due to nitration of the benzene ring present in **phenylalanine, tyrosine, and tryptophan** to give yellow nitro substitution products, which turn orange colored upon the addition of alkali (salt formation). Most proteins give this test positive.
Millon's test: 2 mL of original solution + 2 mL of Millon's reagent (10% mercuric sulfate in 10% H_2SO_4). Boil for 1 minute; cool and add few drops of freshly prepared 1% $NaNO_2$ solution, heat gently.	White precipitate, which turns red on heating	Hydroxyphenyl group (tyrosine) is present. **Principle:** The test is not specific since it is given by all phenolic compounds. Mercuric sulfate and sodium nitrite in acidic medium react with hydroxyphenyl group to form nitrated mercuric phenolate ion, which forms a red-colored complex. **Tyrosine**, the only phenolic amino acid in proteins, gives the test.
Hopkins-Cole (aldehyde) test: 2 mL of original solution + 2 drops of dilute formalin + 2 drops of mercuric sulfate solution, add 2 mL of concentrated H_2SO_4 slowly from the side of the test tube.	Violet ring at the interface of the liquids	Indole group (tryptophan) is present. **Principle:** Sulfuric acid in the presence of mercuric sulfate oxidizes the indole group, which on reacting with formaldehyde forms a violet-colored complex.
Lead acetate test: 2 mL of original solution + 2 mL of 40% NaOH; boil for 2 minutes, then add 3–4 drops of lead acetate. Mix well.	Dark brown/black precipitate	Sulfhydryl group (cysteine, cystine) is present. **Principle:** When protein solution is boiled with strong alkali, sulfur splits off to form Na_2S. Addition of lead acetate results in the formation of brown/black precipitate of lead sulfide. The **test is negative for methionine** as thioether linkage in methionine is strong and sulfur in methionine cannot be split off to form sulfide.
Sakaguchi test: 3 mL of original solution + 1 mL of 40% NaOH + 2 drops of alcoholic α-naphthol + 4–5 drops of freshly prepared sodium hypobromite solution. Mix well.	Intense red color	Guanidino group (arginine) is present. **Principle:** α-naphthol reacts with guanidine group in alkaline medium, which then gets oxidized to form a red-colored complex.
Neumann's test: 2 mL of original solution + 3–4 drops of concentrated HNO_3. Boil it then cool + 2 mL of ammonium molybdate and boil it and cool.	Yellow color precipitates formed	Phosphoprotein is present. **Principle:** On heating with concentrated sulfuric acid and nitric acid, casein is digested and phosphorus is released. Ammonium molybdate reacts with phosphorus in the presence of nitric acid to form a canary yellow precipitate of ammonium phosphomolybdate.

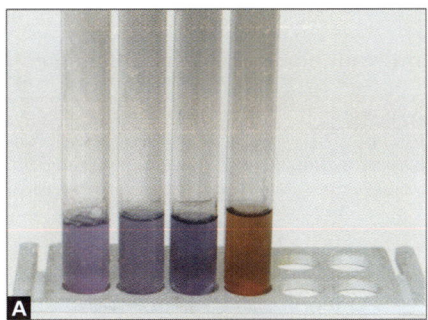

A: Biuret test.

B: Ninhydrin test.

C: Xanthoproteic test.

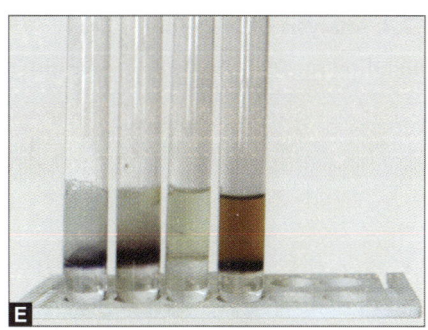

E: Hopkin's Cole's test

F: Lead sulphide test.

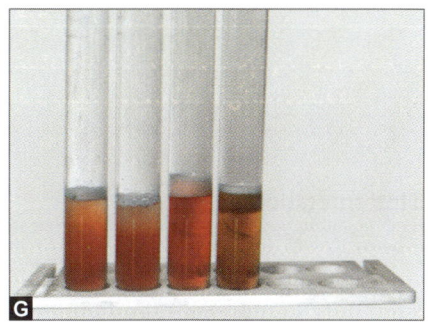

G: Sakaguchi test.

D: Millon Nasse's test.

Figs. 2.2.2A to G: Protein color reactions.
(From left to right Albumin, Casein, Gelatin, Peptone)

Unit 2: Qualitative Experiments and their Clinical Applications

Exercise 2.2.4: Analysis of protein.

Perform the following tests on protein solution provided to you and record your observation and inference.

Test	Observation	Inference

Result and Interpretation

....
....
....
....
....
....
....

Date: **Teacher's signature**

Analysis of protein.

Perform the following tests on protein solution provided to you and record your observation and inference.

Test	Observation	Inference

Result and Interpretation

....
....
....
....
....
....
....

Date: **Teacher's signature**

VIVA VOCE QUESTIONS: PROTEINS

1. What is denaturation of protein?
2. What is the importance of precipitation of proteins?
3. Why proline and hydroxyproline do not give a blue color with Ninhydrin test? If Biuret test is negative but Ninhydrin test is positive in a given solution, what does it indicate?
4. Out of egg albumin, gelatin and casein, which are nutritionally good proteins and why?
5. Can egg albumin is used as an antidote in acute lead or mercury poisoning. Why?
6. Why alcohol is used as a fixative for tissue in histological slides?

R-GROUPS OF PROTEINS AND NUTRITIVE VALUE

1. What is the importance of doing tests for R-groups of proteins and amino acid?
2. Do the R-group tests indicate the nutritive value of proteins?
3. What is a first class protein from nutritional point of view? Give examples.
4. Why methionine does not give the test for sulfur even though it contains sulfur in its R-group?
5. Which test is done to detect a basic amino acid? Name the two basic amino acids.
6. Name a protein which gives most of the R-groups test positive.
7. In Xanthoproteic test, why yellow color changes to orange color?
8. Ordinarily, tyrosine is not an essential amino acid. Under which condition, it becomes an essential amino acid and why?
9. How the presence of histidine can be confirmed by the R-groups test?

NOTES

Analysis of Proteins

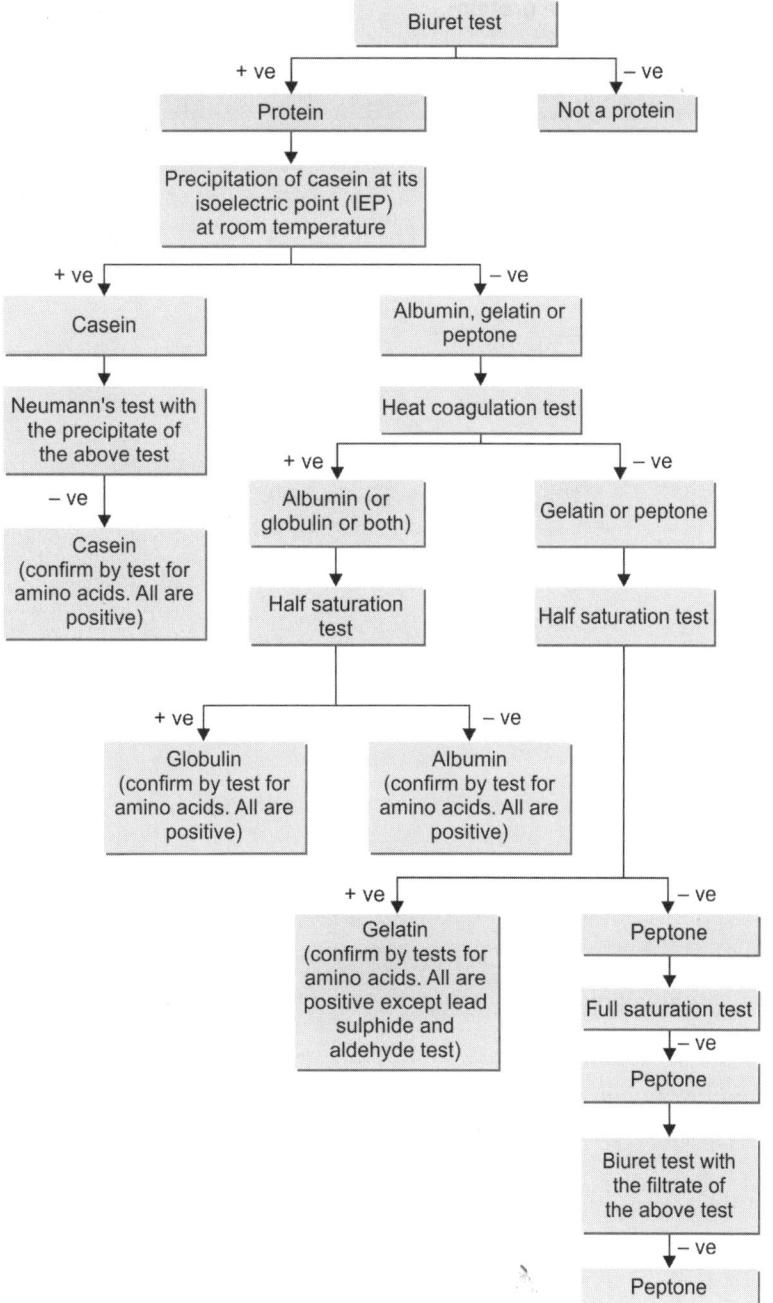

Unit 2: Qualitative Experiments and their Clinical Applications

Exercise 2.2.5: Analysis of protein (unknown protein).

Perform the following tests on protein solution provided to you and record your observation and inference.

Test	Observation	Inference

Result and Interpretation

....
....
....
....
....
....

Date: **Teacher's signature**

Analysis of protein (unknown protein).

Perform the following tests on protein solution provided to you and record your observation and inference.

Test	Observation	Inference

Result and Interpretation

....
....
....
....
....
....
....

Date: **Teacher's signature**

Analysis of Physical and Chemical Composition of Physiological Urine

2.3

> **LEARNING OBJECTIVES**
> At the end of the session the student must be able to:
> ○ Know and identify normal constituents of urine
> ○ Interpret and correlate the finding of normal urine with various physiological conditions
> ○ Perform urine analysis and determine normal urine independently.

Competency Bl.11.3: Describe the chemical components of normal urine.
Competency Bl.11.4: Perform urine analysis to estimate and determine normal constituents.
Domain: Knows.
Level: Knows how.
Core competency: Yes.

URINE ANALYSIS

Urine is an excretory product of the body, which is formed in the kidneys. The composition of urine is a mirror not only of renal function, but also of many physiological and metabolic processes occurring in the body. Thus, examination of urine is quick and convenient method of obtaining information of considerable diagnostic, prognostic, and therapeutic values.

PHYSICAL PROPERTIES OF NORMAL URINE

- *Volume:* Normal urinary volume varies from 600 mml/day to 2,500 mL/day, average—1,500 mL/day. Urine volume is influenced by fluid intake, fluid loss, cardiovascular, and renal function.
- *Appearance:* Normal fresh urine is clear. Turbidity may develop on standing as the pH increases and the phosphates precipitate.
- *Color:* Normal freshly voided urine is colorless to straw colored. This is due to the presence of pigment **urochrome**. Traces of other substances that can contribute are uroerythrin, urobilin, uroporphyrin, coproporphyrin, and uroresin. The color may be light or dark depending on the volume.
- *Odor:* Fresh urine has an aromatic odor or ammoniacal smell due to volatile organic acids. Some foods like onion and garlic impart their smell to urine.
- *pH:* Normally, pH is 4.5–8.0 with a mean of 6.0 in freshly voided urine. It is measured by litmus paper. Person with high on protein diets, the urine is more acidic because more sulfates and phosphates are eliminated from the protein catabolism. Urine on standing becomes alkaline by the bacterial action on urea and formation of ammonia. After meals, due to HCl secretion in the stomach, the urine becomes alkaline due to the **alkaline tide**.
- *Specific gravity:* The specific gravity of physiological urine lies in the range of 1.012–1.025. It indicates the concentrating ability of the kidney. The specific gravity is dependent on the concentration of solutes in the urine. It is measured by **urinometer**. This instrument floats in the urine, the calibration mark, which corresponds

to the surface level of the urine, is read. The lower the specific gravity, the further the urinometer sinks.

CHEMICAL COMPOSITION OF NORMAL URINE

Normal urine contains 90–95% water and rest comprises of the solid, which may be organic or inorganic.

Major organic constituents excreted are:

Urea	- 25–30 g/day
Uric acid	- 0.5–0.8 g/day
Creatinine	- 1–1.8 g/day

Inorganic constituents are:

Chlorides	- 10–15 g/day
Sodium	- 3.5 g/day
Potassium	- 2–2.5 g/day
Calcium	- 0.1–0.3 g/day
Phosphates	- 0.8–1.3 g/day
Sulfates	- 1–1.2 g/day
Ammonia	- 0.7–0.8 g/day

URINE COLLECTION AND PRESERVATION FOR ANALYSIS

The urine specimen can be collected in following ways:

- **Fresh urine sample:** Morning sample is preferred for investigations. After getting up from the bed say at 6.30–7 AM, the midstream sample (initial part is voided and then collected followed by last part voided) is collected in a clean bottle. At least 10–20 mL is adequate. For bacteriological examination, collection should be done in a sterile bottle.
  Uses: For routine investigation.
  Physical examination: Appearance, pH, color, odor, specific gravity, volume, etc.
  Biochemical examination: Sugar, protein, ketone bodies, blood, bile salts and bile pigments, urobilinogen, and others.
  Microscopic examination: Casts, crystals (like calcium oxalate, urate), RBCs, pus cells, bacteria, etc.
- **24-hour urine collection:** This is done by asking the patient to void in the morning, say at 7:00 AM after getting up from bed. This sample is discarded. All subsequent samples passed during the day and night, and next day 7:00 AM. Sample after getting up from bed is collected in a clean bottle or can with or without preservative as per the directive from laboratory. The volume is measured, mixed, and a small amount as per direction is sent for examination.
  Uses:
  - Urinary protein estimation.
  - Specific gravity
  - Measuring clearances, GFR measurement
  - Hormone assays
  - Mineral estimation, viz. calcium.

  Preservatives used in 24-hour urine collection:
  - Thymol (0.1 mg in 100 mL urine) for protein and creatinine
  - Formalin (40%) or hydrochloric acid (two drops/30 mL of urine) for calcium, amino acids, and catecholamines
  - Toluene or chloroform (50 drops/24 h urine).

3. **Random sample:** Taken any time during the day.
   Uses: Mainly for sugar and ketone bodies.

TESTS FOR NORMAL CONSTITUENTS OF URINE

Test for Calcium

Take ammonium oxalate solution in a test tube. Add small amount of urine 1 mL. A white ppt. of calcium oxalate is obtained.

Test for Phosphate

Take urine sample and concentrated HNO_3 in a test tube and heat it. Add ammonium molybdate solution. A canary yellow ppt. shows the presence of phosphate.

Test for Urea

i. **Specific urease test:**
   Reagents:
   1. Soybean meal/jack bean meal
   2. Phenol red/phenolphthalein indicator.

   Procedure: Take 5 mL of urine in a test tube and small amount of soybean meal and 1–2 drops of phenol red/phenolphthalein indicator. Mix and allow to stand for 15 minutes. Yellow color changes to red due to evolution of ammonia from urea.

   $$CO(NH_2)_2 + H_2O \xrightarrow{\text{Urease}} CO_2 + 2NH_3$$

ii. **Sodium hypobromite test:**
    Reagent: Alkaline hypobromite.

Unit 2: Qualitative Experiments and their Clinical Applications

Table 2.3.1: Physical examination of normal urine.

Appearance	Freshly void urine is clear, it becomes turbid on standing due to precipitation of phosphates
Odor	Aromatic odor, which becomes ammonical on standing
Color	Straw color
Volume	800 mL–2 L/day, depends upon the intake of water
pH	Usually acidic, but it ranges from 4.5 to 8
Specific gravity	1.010–1.025, measured with the help of urinometer, it tell us about the concentrating ability of kidneys

Source: Harrison's Principles of Internal Medicine, 20th Edition, McGraw-Hill Publisher, 2018.

Procedure: Take 5 mL of urine in a test tube. Add 1 mL of alkaline sodium hypobromite. Mix gently. A marked effervescence occurs due to evolution of nitrogen from urea. CO_2 will be absorbed NaOH.

$$CO(NH_2)_2 + 3\,NaOBr + 2\,NaOH \longrightarrow NaBr + N_2 + Na_2CO_3 + 3H_2O$$

Test for Creatinine

Jaffe's test: Take 3 mL of urine in test tube. Add 1 mL of alkaline picrate. A orange red color shows creatinine due to formation of creatinine picrate.

Table 2.3.2: Analysis of normal constituent of physiological urine.

Test	Principle	Observation	Inference
Urea **1. Specific urease test reagents:** Soybean meal, phenol red **Procedure:** 4–5 mL of urine + pinch of soybean meal + 1–2 drop of phenol indicator mix and wait 10 mins	Urease in soybean meal acts on urea and liberate NH_3, which makes the medium alkaline. In alkaline medium phenolphthalein gives red color	Yellow to red color (Due to NH_3)	Presence of urea
2. Sodium hypobromite test reagents: Alkaline hypobromite **Procedure:** 4–5 mL sample + 2 mL alkaline hypobromite + Mix	Breakdown of urea and release of nitrogen produce effervescence	Effervescence due to production of nitrogen	Presence of urea
Calcium test **Reagents:** Ammonium oxalate **Procedure:** Take 2–3 mL ammonium oxalate + 1–2 mL urine	Formation of calcium oxalate	White precipitate of calcium oxalate	Presence of calcium
Phosphate test **Reagents:** Ammonium molybdate **Procedure:** Take 1–2 mL urine + add con. HNO_3 + heat + ammonium molybdate	Formation of phospho-molybdate	Canary yellow precipitate	Presence of phosphate
Creatinine test **Jaffe's test** **Reagents:** Alkaline picrate **Procedure:** 3–4 mL urine + 1 mL alkaline picrate	Formation of creatinine picrate	Orange red color	Presence of creatinine

Source: Simplified Practical Manual of Biochemistry, 2nd Edition, Jaypee Publishers, New Delhi, 2018.

VIVA VOCE QUESTIONS: PHYSIOLOGICAL URINE

1. What is polyuria? What are its common causes? What is the characteristic feature of polyuria seen in diabetes mellitus?
2. What are common causes of oliguria and anuria?
3. What is isosthenuria and what are its causes?
4. What is normal specific gravity of urine? What are the common causes of increase in specific gravity?
5. Mention some common abnormalities of change in urine color and their significance.

NOTES

Analysis of Physical and Chemical Composition of Physiological Urine

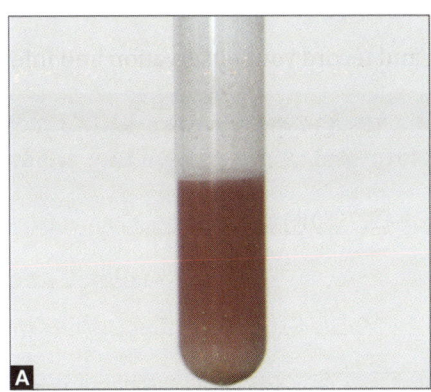 

Figs. 2.3.1A and B: Test for Urea: A. Urease test/Soyaben meal test; B. Sodium Hypobromite test.

 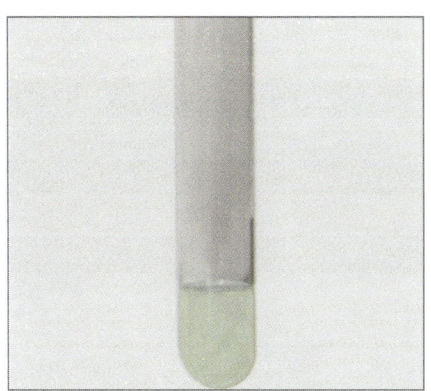

Fig. 2.3.2: Test for phosphorus. **Fig. 2.3.3:** Test for calcium.

Fig. 2.3.4: Test for creatinine **Fig. 2.3.5:** Full pic analysis of physiological urine.

From left to right

Urease test, Sodium Hypobromite test, Test for Phosphorus, Test for calcium, Test for creatinine.

Unit 2: Qualitative Experiments and their Clinical Applications

Exercise 2.3.1: Analysis of physiological urine sample.

Perform the following tests on protein solution provided to you and record your observation and inference.

Test	Observation	Inference

Result and Interpretation

....
....
....
....
....
....
....

Date: **Teacher's signature**

Analysis of Physical and Chemical Composition of Physiological Urine

Analysis of physiological urine sample.

Perform the following tests on protein solution provided to you and record your observation and inference.

Test	Observation	Inference

Result and Interpretation

....
....
....
....
....
....
....

Date: **Teacher's signature**

Identify, Perform and Interpret Pathological Urine Analysis and Correlate it with Pathological States

2.4

LEARNING OBJECTIVES

At the end of the session the student must be able to:
- Know and identify abnormal constituents of pathological urine
- Interpret the finding of abnormal urine with pathological conditions
- Perform urine analysis and correlate results with pathological status
- Perform and interpret urinary sugar and ketone body estimation with a dipstick.

Competency BI.11.4: Perform urine analysis to estimate and determine normal and abnormal constituents.
Competency BI.11.20: Identify abnormal constituents in urine, interpret the finding, and correlate these with pathological states.
Competency PE.21.11: (Integration with pediatrics) Perform and interpret the common analytes in a urine examination.
Competency PE.33.6: (Integration with pediatrics) Perform and interpret urine dipstick for sugar.
Competency IM.11.13: (Integration with Internal Medicine) Perform and interpret a urinary ketone estimation with a dipstick.
Domain: Knows.
Level: Knows how.
Core competency: Yes.

URINE ANALYSIS IN PATHOLOGICAL CONDITIONS

Urine can be investigated under the following headings:
1. Physical examination
2. Biochemical examination
3. Microscopic examination
4. Bacteriological examination.

However, (1) and (2) are important for us. The urine is said to be pathological when it contains **abnormal constituents** in detectable amounts. The abnormal constituents commonly tested in biochemistry laboratory are proteins, sugars, ketone bodies, bile salts and pigments, blood, urobilinogen, and other constituents excreted in specific disease condition.

PHYSICAL EXAMINATION

1. **Appearance:** If the freshly voided urine is turbid, it may be due to:
   - Presence of pus cells in urinary tract infection
   - Presence of fat globules as in chyluria or lipiduria.
2. **COLOR:**
   - **Deep yellow:** Due to bile pigments in jaundice.
   - **Reddish:** Due to red cells (hematuria) in urinary stones, cancer, urinary schistosomiasis, and trauma.
   - **Brownish:** (Hemoglobinuria) due to blackwater fever, myoglobinuria, and porphyrias (congenital erythropoietic porphyria).
   - **Brown to black:** In alkaptonuria, fresh urine is of normal color and on standing, it changes to brown, methemoglobinuria, phenol poisoning.
   - **Dark amber:** Concentrated urine or vitamin B-complex therapy.
   - **Pale:** Dilute urine in diabetes insipidus or polyuria.
   - **Orange:** Due to certain drugs like rifampicin or riboflavin.
3. **Volume:** Changes in 24-hour urine are seen in following conditions:

- **Polyuria:** Volume more than 3000 mL/day due to:
  - Hyperosmolar polyuria as in diabetes mellitus
  - Hypoosmolar polyuria in diabetes insipidus.
- **Isosthenuria:** (Watery urine) in chronic renal failure.
  - Other causes include—compulsive polydipsia, alcohol ingestion, nervous disorder, and drugs intake, viz. diuretics, lithium.
- **Oliguria:** Volume less than 500 mL/day due to:
  - Severe dehydration, acute renal failure
  - Reduced circulatory fluid volume, viz. edema, fever
  - Loss of fluids as in vomiting, diarrhea.
- **Anuria:** Complete cessation of urine. Usually, it means severe oliguria (<50 mL/day) as in:
  - Acute tubular necrosis (kidney failure)
  - Post blood transfusion reaction
  - Postsurgical shock.

4. **Specific gravity:** In most conditions, whether it is physiological or pathological, specific gravity of urine is inversely proportional to the volume of urine, except diabetes mellitus.
   - SG < 1.016: Hypo-osmolar condition seen in diabetes insipidus, compulsive polydipsia.
   - SG > 1.025: Hyperosmolar condition seen in diabetes mellitus (presence of glucose)
     - Severe dehydration
     - Nephrotic syndrome (presence of protein in excess)
     - Adrenal insufficiency (presence of excess NaCl).
   - SG = 1.010: **Isosthenuria**.
     Fixed specific gravity is seen in CRF (chronic renal failure).

   Correction of specific gravity: Urinometer is standardized at 15°C. So, for every 3° rise or fall of temperature, we add or subtract 0.001 from original reading, respectively.

   In case of proteinuria, true specific gravity can be determined mathematically by deducting 0.003 from the observed value for every gram of albumin present per 100 mL of urine.

5. **pH: Decrease in urinary pH:**
   - Metabolic acidosis
   - Diabetic ketoacidosis.

   Increase in urinary pH:
   - Metabolic alkalosis observed in prolonged vomiting
   - Urinary tract infection.

6. **Odor:**
   - Odor due to ingested foods like garlic is not pathological
   - Smell of acetone in diabetic ketoacidosis, chronic starvation
   - Foul smell due to bacterial infections
   - Mousy smell—phenylketonuria.

BIOCHEMICAL EXAMINATION

Proteins

The presence of proteins in urine is usually pathological and is known as proteinuria. The protein commonly found in urine is **albumin**, hence proteinuria is synonymously used as albuminuria. The most common causes are divided into:

i. **Prerenal:**
   - Heart failure
   - Fever
   - Toxemia of pregnancy.

ii. **Renal:**
   - Nephrotic syndrome
   - Inflammatory kidney disease's
   - UTI.

iii. **Postrenal:** Stones in urinary tract
   - UTI.

Note: Nephrotic syndrome is the most common cause. Sometimes, there is **transient** albuminuria which is physiological:
- Due to severe exercise in unaccustomed individual
- Due to postural changes, viz. long period of standing.

Note: Microalbuminuria—excretion (30–300 mg/day) of albumin is called microalbuminuria. It appears to be predictor of future development of clinical renal disease in patients with HT or diabetes mellitus.

TESTS FOR PROTEIN

Heat Coagulation Test

Fill three-fourths of the test tube with urine. Heat the upper portion. Lower part serves as control. Note for the appearance of turbidity. A coagulum indicates presence of proteins. Now, add 2% of acetic acid (few drops). If the coagulum dissolves, it is due to the presence of excess phosphates (phosphaturia). If it does not dissolve, it is due to proteins.

This is the most definitive test for urinary proteins because albumin is heat coagulable.

Observation		Interpretation
Mild	+	Visible turbidity
Moderate	2+	Moderate turbidity
Severe	>2+	Heavy turbidity

Albustix, a dry chemistry technique, is also available to detect albuminuria by dipping the strip in the urine and checking the changes even by the patient himself **(Fig. 2.4.1)**.

Heller's Test

Take 3 mL of nitric acid (concentrated) in a test tube. Add 3 mL of urine along the sides of the test tube. A white ring of

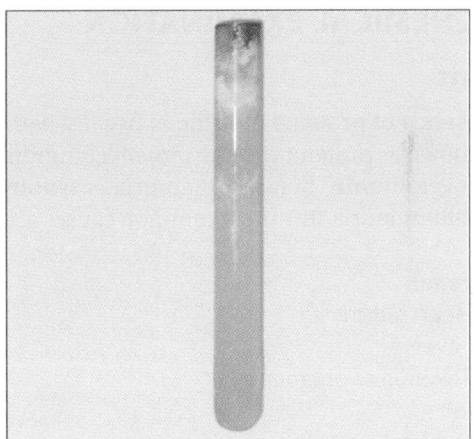

Fig. 2.4.1: Positive heat coagulation test for proteins in pathological urine.

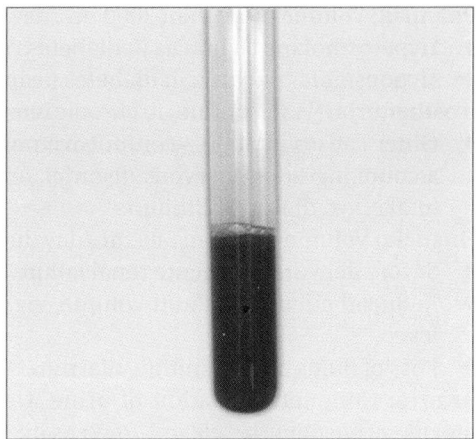

Fig. 2.4.2: Positive Benedict's test for reducing sugars in pathological urine.

acid metaproteins at the junction of the two fluids indicates protein is present.

Principle: Nitric acid causes precipitation of protein.

BENCE-JONES PROTEINS

This protein is seen in urine of **multiple myeloma patients**. These are light chain components of immunoglobulins (MW: 16,000–18,000 kDa).

This protein has a characteristic property to precipitate between 40°C and 60°C on heating the urine sample and dissolves on further heating to 100°C. The precipitate reappears again on cooling between 40°C and 60°C.

A special band M-band (myeloma band) appears on serum protein electrophoresis for proteins in these patients.

TEST FOR REDUCING SUGARS

Benedict's Qualitative Test

Take 5 mL of Benedict's qualitative reagent in a test tube. Add 8 drops of urine. Boil for 2 minutes and cool. A change from blue to green, yellow, orange, or red indicates the presence of reducing sugar in the urine **(Fig. 2.4.2)**.

Note: While checking the color, mix the solution and see the color of the solution to interpret the result.

Composition of Benedict's Qualitative Reagent

$CuSO_4$ (provides cupric ions)
Sodium carbonate (provides alkaline medium)
Sodium citrate (keeps cupric ions in solution)

Principle

In hot alkaline medium, reducing sugar is converted to enediol form, which has strong reducing properties and reduces $CuSO_4$ to cuprous hydroxide that decomposes to cuprous oxide (red ppt.). This test is a semiquantitative in nature, so helps in assaying the extent of sugar excretion in urine.

Observation		Interpretation
Green precipitation	0.5–1 g/100 mL	+
Yellow precipitation	1–1.5 g/100 mL	2+
Orange precipitation	1.5–2 g/100 mL	3+
Brick red precipitation	2 g/100 mL	4+

Some nonsugar reducing substances may give false-positive result
- Creatinine, ascorbic acid, urates, and glucoronates
- Drugs like salicylates, PAS, and isoniazid.

Various reducing sugars that may give positive test are:
- **Glucose (Glucosuria):** Most commonly seen in diabetes mellitus, glucosuria, renal alimentary glucosuria, and Fanconi syndrome. Renal glycosuria is a condition in which threshold for glucose (180 mg%) is decreased to 150 mg%. So, glucose appears in urine at this level also.
- **Lactose (Lactosuria):** Lactating mother, last trimester of pregnancy.
- **Fructose (Fructosuria):** Essential fructosuria, hereditary fructose intolerance.
- **Galactose (Galactosuria):** Galactosemias in children.
- **Pentoses (Pentosuria):** Sometime, high intake of fruits like cherry, plums due to deficiency of xylitol dehydrogenase enzyme.

If reducing sugar is detected in urine, blood sugar level must be correlated with the findings. Nowadays, presence of glucose in urine can be confirmed by dry chemistry method using Glucostix, i.e. a strip impregnated with glucose oxidase reagent, which can be performed even by the patient him or herself at the residence.

Note: The paper chromatography technique is usually performed in the laboratory to identify the specific nature of the sugar in the urine, if it is required to confirm the diagnosis.

TEST FOR KETONE BODIES (ROTHERA'S NITROPRUSSIDE TEST)

Saturate 5 mL of urine with solid $(NH_4)_2SO_4$ and add 0.5 mL of freshly prepared sodium nitroprusside solution. Mix well and add equal volume of liquor ammonia from the side of the test tube. A purple ring at the junction of the liquids indicates the presence of ketone bodies **(Fig. 2.4.3)**.

It is given by acetoacetate and acetone, but not by β-OH butyrate. False positive if urine contains or has L-dopa or phenylpyruvate. Dipsticks (Ketostix on dry chemistry principle) are also available for detection. Presence of ketone bodies indicates:
- Severe uncontrolled diabetes mellitus (diabetic ketoacidosis)
- Starvation
- Toxemia of pregnancy
- Intake of high fat and low carbohydrate diet.

BILE SALTS

This test in urine is performed to assess the liver function. Bile salts are sodium and potassium salts of glycocholic and taurocholic acids formed in the liver from cholesterol. They are secreted in bile. They reduce surface tension and aid in digestion and absorption of fats in intestine.

Hay's Sulfur Test

Sprinkle a little dry sulfur powder on the surface of fresh urine in a test tube taking **distilled water as control.** Do not shake or disturb the tube. If the sulfur powder sinks, bile salts are present. Urine preservative (thymol) gives false test **(Fig. 2.4.4)**.

BILE PIGMENTS

Presence of bilirubin in urine means excretion of **conjugated bilirubin (bilirubin glucuronide)**, usually seen in hepatic and obstructive jaundice. Urine containing

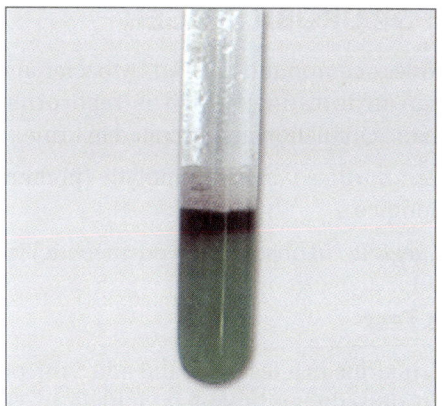

Fig. 2.4.3: Positive Rothera's test for ketone bodies in pathological urine.

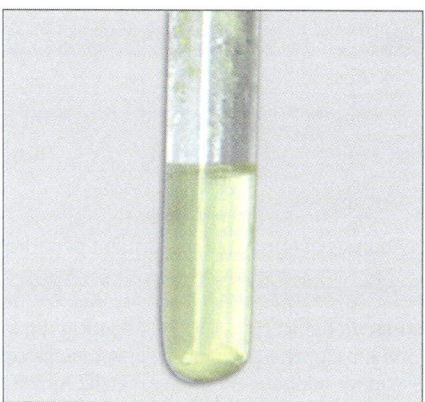

Fig. 2.4.4: Positive Hay's sulfur test for bile salts in pathological urine.

bilirubin typically has a beer brown color when voided and produces yellow form when shaken (fresh sample must be examined before bilirubin gets converted to biliverdin).

Fouchet's Test

$BaCl_2$ reacts with sulfur radicals in urine to form barium sulfate. If bilirubin is present in urine, it adheres to the ppt. and is detected by oxidation of bilirubin to biliverdin.

Add 5 mL of 10% $BaCl_2$ to 10 mL of urine and filter. Dry the filter paper and add a few drops of Fouchet's reagent (prepared by adding 10 mL of 10% $FeCl_3$ to 100 mL of 25% TCA). A greenish-blue color denotes the presence of bilirubin (due to oxidation of bilirubin to biliverdin conversion produce green color, biliverdin to bilicyanin conversion produce blue color by $FeCl_3$).

TEST FOR UROBILINOGEN

It is a colorless compound derived by bacterial oxidation of bilirubin in intestine, which is reabsorbed by the enterohepatic circulation and excreted in urine.

Ehrlich's test positive means: Hemolytic (prehepatic) and hepatic jaundice.

The test is negative in obstructive (posthepatic) jaundice.

Ehrlich's Test

Take 5 mL of urine in a test tube and add 5 mL of Ehrlich's reagent (p-dimethylaminobenzaldehyde in HCl) to it. Wait for 10 minutes and add 10 mL of saturated sodium acetate solution. A pinkish-red color indicates presence of urobilinogen.

	Hemolytic	Hepatocellular/ Hepatic	Obstructive
Bilirubin	Absent	+	++
Bile salts	Absent	+	++
Urobilinogen	++	+/Absent	Absent

TEST FOR BLOOD PIGMENT

Presence of RBC in urine is called hematuria.

Benzidine Test

It is used for blood as well as hemoglobin in urine.

To 1–2 drops of saturated benzidine solution, add 2 mL of urine and 3–4 drops of H_2O_2. A blue or green color developing indicates the presence of blood. H_2O_2 liberates nascent oxygen that oxidizes benzidine to a colored derivative.

Benzidine is carcinogenic chemical.

Causes

Gross hematuria is found in renal stones.
- Malignancies, tuberculosis, and trauma of kidney
- Acute glomerulonephritis.

Microscopic hematuria may be seen in:
- Malignant hypertension
- Sickle cell anemia
- Coagulation abnormalities
- Polycystic kidney diseases.

Table 2.4.1: Analysis of abnormal constituent of pathological urine.

Test	Principle	Observation		Inference
Reducing sugar: Benedict test **Reagents:** Sodium citrate + sodium carbonate + copper sulphate **Procedure:** 5 mL reagent + 7–8 drop of urine sample + boil 2–3 min + cool	Reducing sugars have free aldehyde or ketone group which reduces metallic ions. Benedict reagent has cupric ion which is reduced to cuprous ion	**Color** Blue Green Yellow Orange Brick red	**Concentration** No sugar 0.5 -1 1-1.5 1.5-2 >2	Presence of reducing sugar
Protein heat coagulation test- procedure: Fill ¾ test tube with sample + heat upper part + coagulum formed + add 2–3% CH_3COOH if coagulum is dissolves, indicate presence of phosphate	Denaturation of protein by heat	Coagulum is formed at upper part		Presence of protein
Heller's test-reagents: HNO_3 **Procedure:** 3 mL of sample + add 3–4 mL of HNO_3 along the side wall of tube	Nitric acid cause precipitation of proteins	White ring formed at the junction of two liquid		Presence of protein
Bence Jones proteins: **Procedure:** Heat 40–60°C– ppt + heat 100°C– dissolve ppt + cool again reappear		Precipitate dissolve by heating and reappear on cooling		Presence of Bence Jones protein
Ketone bodies: Rothera's nitroprusside test reagents: Ammonium sulfate, sodium nitroprusside, liquor ammonia **Procedure:** 5 mL urine saturate with ammonium sulfate + sodium nitroprusside + shake + liquor ammonia side of tube	Ketone bodies form a purple colored complex with sodium nitroprusside in alkaline medium. Hydroxybutyrate does not	Purple ring formed at the junction of two liquid		Presence of ketone bodies

Contd...

Contd...

Test	Principle	Observation	Inference
Bile salt Hay's sulfur test reagents—sulfur powder **Procedure:** 4–5 mL of sample in tube + add pinch of sulfur powder on surface	Bile salt reduces the surface tension of water	Sulfur powder sinks at bottom	Presence of bile salt
Bile pigments Fouchet's test reagent: (FeCl$_3$+TCA) 6 mL BaCl$_2$ add 10–11 mL urine sample + filter Add 1–2 drop of reagent	BaCl$_2$ reacts with sulfur to form barium sulfate	Green blue color appear	Presence of bilirubin
Urobilinogen Ehrlich's test- reagent: (p-dimethyl- aminobenzaldehyde) **Procedure:** 5–6 mL urine + 5 mL of reagent + wait 8–10 min + add 10 mL saturated sodium acetate	Urobilinogen in acidic medium reacts with Ehrlich's reagent and gives color compound	Pink, red color	Presence of urobilinogen
Blood pigment benzidine test- reagents: H$_2$O$_2$ **Procedure:** 2–3 drop saturate benzidine solution + add 3 mL sample + 4 drop H$_2$O$_2$	Peroxidase like activity of hemoglobin H$_2$O$_2$—H$_2$O + O (O) oxidizes the benzidine to form blue/green colored oxidized product	Blue green color develop	Presence of blood pigment

Source: An Easy Guide for Practical Biochemistry, 2nd Edition, Jaypee Publishers, 2018.

Principle: Based upon peroxidase-like activity of Hb, which breaks up the H$_2$O$_2$ into water and nascent O$_2$.

$$H_2O_2 \xrightarrow{Peroxidase} H_2O + [O]$$

[O] oxidizes the benzidine to form a blue/green-colored oxidation product.

VIVA VOCE QUESTIONS: PATHOLOGICAL URINE

1. What is glycosuria? What are its causes? How would you distinguish between a pathological and physiological glycosuria?
2. Why glacial acetic acid is added if there is turbidity in heat coagulation test for protein detection?
3. Which protein is most commonly excreted in earliest stages of glomerular damage and why?
4. Can you suspect multiple myeloma by doing a simple heat coagulation test of urine? What is "M" band? What is its cause?
5. How ketone bodies are detected in urine? Are they produced normally? What is their biochemical significance?
6. Discuss clinical condition in which bile acids and bile salts are present in urine.
7. What are the precautions to be taken in performing Hay's sulfur test?

NOTES

Identify, Perform and Interpret Pathological Urine Analysis and Correlate it with Pathological States

Exercise 2.4.1: Analysis of pathological urine.

Perform the following tests on protein solution provided to you and record your observation and inference.

Test	Observation	Inference

Result and Interpretation

....
....
....
....
....
....
....

Date: **Teacher's signature**

Unit 2: Qualitative Experiments and their Clinical Applications

Analysis of pathological urine.

Perform the following tests on protein solution provided to you and record your observation and inference.

Test	Observation	Inference

Result and Interpretation

....
....
....
....
....
....

Date: **Teacher's signature**

UNIT 3

Quantitative Experiments and their Clinical Interpretation

OUTLINES

3.1. Principle of Colorimetry
3.2. Principle of Spectrophotometry
3.3. Estimation of Blood Glucose
3.4. Glucose Tolerance Test and Glycated Hemoglobin
3.5. Liver Function Test
3.6. Kidney Function Test
3.7. Lipid Profile (Atherogenic Profile)
3.8. Estimation of Serum Calcium and Serum Phosphorus

Principle of Colorimetry

3.1

> **LEARNING OBJECTIVE**
> At the end of the session the student must be able to understand the basic principle of colorimetry.

Competency BI.11.6: Describe the principle of colorimetry.
Domain: Knows.
Level: Knows how.
Core competency: yes.

COLORIMETRY

The term colorimetry means measurement of color. This technique is used very commonly in biochemistry to measure the concentration of substances that are colored or that can be converted into colored compounds by suitable reaction. The technique is very sensitive and requires a very small amount of the sample.

Principle: When white light passes through a colored solution, a portion of the light is absorbed by the coloring substance (chromogen), and the rest is transmitted. The extent to which the different components of white light are absorbed by a solution depends upon the nature of the chromogen. Thus, if a chromogen forms a red solution, it means that the absorption of the red component is the minimum whereas the other components are absorbed to a larger extent. The complementary color (blue in this case) is absorbed to the greatest extent.

In colorimetry, it is preferable to use monochromatic light of a specific wavelength, which is absorbed maximally by the chromogen being measured. The relationship between the absorption (or the transmission) of the light and the concentration of the chromogen was described by Beer, and that between the absorption of the light and the thickness of the solution was described by Lambert.

Beer's law: This states that the log of the ratio of intensities of incident light (I_o) and emergent light (I) is directly proportional to the concentration of the coloring substance (C) in the solution provided that the thickness of the solution through which the light passes is constant (K).

$$\log \frac{I_o}{I} = KC$$

Lambert's law: This law states that the log of the ratio of intensities of incident light (I_o) and emergent light (I) is directly proportional to the thickness of the solution (t) through which the light passes provided that the concentration of the chromogen is constant.

$$\log \frac{I_o}{I} = K2t$$

Beer-Lambert's law is:

$$\log \frac{I_o}{I} = KCt$$

Thus, the Beer-Lambert's law states that the log of ratio of the intensities of incident light and emergent light is directly proportional to the concentration of the chromogen and the thickness of solution through which the light passes.

The ratio of the intensities of emergent light and incident light (I/I_o) is known as transmittance (T). This is a measure of the ability of a solution to transmit light.

So,

$$\log \frac{1}{T} = KCt$$

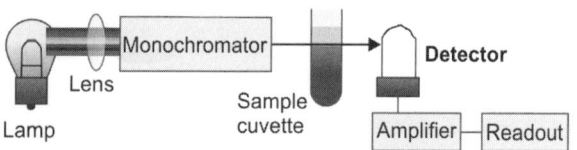

Fig. 3.1.1: Principle and components of colorimeter.

log 1/T is known as the optical density (OD) or the absorbance (A). This is a measure of the ability of a solution to absorb light.

So, A = KCt

and A ∝ Ct

In other words, whereas transmittance (T) has a logarithmic relationship with C and t, absorbance (A) has a direct relationship with C and t. Therefore, absorbance is used in calculations in all the colorimetric determinations **(Fig. 3.1.1)**.

PHOTOELECTRIC COLORIMETER

Photocolorimeter measures the intensity of transmitted light through a color solution. The concentration of colorless biochemical compounds and metabolites can be estimated if they are converted into colored compounds **(Fig. 3.1.2)**.

The basic components of a photoelectric colorimeter are:

- **Source of light:** An electric lamp, e.g. tungsten lamp issued as a source of light. The intensity of light varies in different instruments **(Table 3.1.1)**.
- **Filters:** The light from the electric lamp is passed through a filter. The filters are usually made up of colored glass or dyed gelatin. A set of filters is provided with the instrument. For each determination, a suitable filter should be selected. In some instruments, a prism or diffraction grating is used instead of filters. This disperses the white light into its components. By passing the dispersed light through a narrow slit, monochromatic light of a specific wavelength can be obtained; such instruments are known as spectrophotometer.

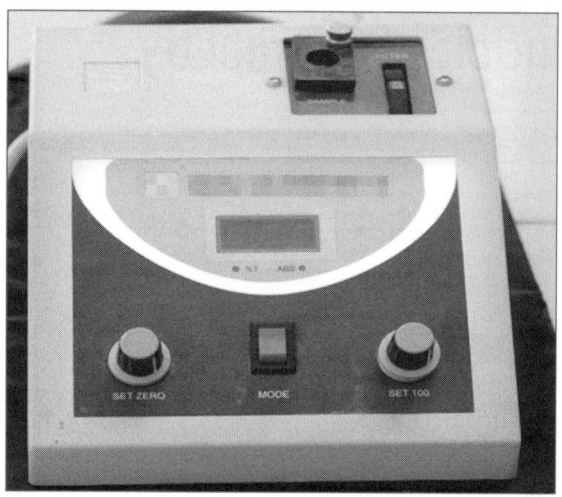

Fig. 3.1.2: Colorimeter.

Table 3.1.1: UV, visible, and infrared spectrum characteristics.

Wavelength (nm)	Region name	Color observed/ complementary color (absorbed/reflected)
<380	UV	Not visible
380–440	Visible	Violet (green yellow)
440–500	Visible	Blue (yellow)
500–580	Visible	Green (red)
580–600	Visible	Yellow (blue)
600–620	Visible	Orange (green blue)
620–750	Visible	Red (green)
>750	IR	Not visible

- **Cuvettes:** These are glass tubes of usually 1 cm diameter and uniform thickness.
- **Photocell or phototube:** It is used to convert the transmitted light energy into electrical energy.
- **Measuring device:** The potential difference generated in photocell or phototube is measured by a sensitive galvanometer. The galvanometer is generally calibrated to read directly either transmittance or absorbance or both.

Exercise 3.1.1: Verification of Lambert-Beer's law.

Take six test tubes and label them as blank (B) and tests. Prepare standard solution of albumin (1 mg/mL). Add distilled water and Biuret reagent in given scheme below.

Test tubes	Albumin (standard)	Distilled water	Biuret reagent
Blank	0.0 mL	2.5 mL	3–5 mL
T1	0.5 mL	2.0 mL	3–5 mL
T2	1.0 mL	1.5 mL	3–5 mL
T3	1.5 mL	1.0 mL	–
T4	2.0 mL	0.5 mL	3–5 mL
T5	2.5 mL	0.0 mL	3–5 mL

Mix and wait for 10 minutes and measure absorbance of each solution at wavelength 540 nm setting zero blank. Plot a graph taking albumin concentration on X-axis and OD on Y-axis, it should give linear curve.

Observation and Calculation

Test tubes	OD values
T1	
T2	
T3	
T4	
T5	

Draw a graph with the help of above finding.

Result and Interpretation

....
....
....
....
....
....

VIVA VOCE QUESTIONS: COLORIMETER

1. What is transmittance and absorbance? How are they related?
2. The OD that is measured in colorimeter, is it absorbance or transmittance? What is its range?
3. What do you understand by absorption maxima?
4. How do you define molecular extinction coefficient or absorption coefficient of a substance? What is its importance?
5. What are the qualities of a good cuvette?
6. How can a concentrated test sample be analyzed by colorimetry when it does not obey the Lambert-Beer's law?
7. What do you understand by a standard solution? What is the importance of a "blank"?
8. For photometric estimations, is it necessary to always have a colored solution?
9. What are the advantages of spectrophotometer over colorimeter?
10. What are the sources of light used in colorimeter and spectrophotometer?

NOTES

Principle of Spectrophotometry

3.2

> **LEARNING OBJECTIVE**
> At the end of session student must be able to understand the basic principle of spectrophotometry.

Competency BI.11.18: Discuss the principle of spectrophotometry.
Domain: Knows.
Level: Knows how.
Core competency: Yes.

INTRODUCTION

When light passes through or is reflected from a sample, the amount of light absorbed is the difference between the incident radiation, I_o, and the transmitted radiation, I. The amount of light that is absorbed is expressed either as transmittance or absorbance. Transmittance is usually given in terms of a fraction of 1 or as a % and is defined as follows (**Fig. 3.2.1**):

$$T = I/I_o = 10^{-K'C}$$

Absorbance is defined as follows:

$$A = \log I_o/I = K''C$$

Fig. 3.2.1: Spectrophotometer.

INSTRUMENTATION

General

In principle, all spectrophotometers consist of four major subunits:
i. A source that generates electromagnetic radiation
ii. A dispersion device that selects a particular wavelength (correctly a waveband) from the broadband radiation of the source
iii. A sample area
iv. A detector to measure the intensity of radiation.

In addition, there are other optical components such as lenses or mirrors that relay the light through the instrument. To determine the degree of interaction of the sample with radiation (transmittance or absorbance), it is necessary to measure both the intensity of the incident radiation, i.e. without the sample, and the transmitted intensity, i.e. with the sample.

Sources

The ideal source would give a constant intensity over all wavelengths in the UV-visible region with low noise and long-term stability. This ideal source does not exist. Two sources are commonly used:
1. The deuterium arc lamp gives good intensity continuum in the UV region with useful intensity in the visible region. Modern lamps have low noise characteristic, although the noise from the lamp is often the limiting

factor in overall instrument noise performance. Over time, the intensity of light from a deuterium arc lamp decreases steadily. Typically, a lamp will have a "half-life" (the time for the intensity to fall to half the initial value) of about 1,000 hours.

2. The tungsten-halogen lamp gives good intensity over the part of the UV and the whole of the visible range. It has very low noise and low drift. Its useful life is typically 10,000 hours.

Most spectrophotometers that cover the UV-visible range include both of these lamps and some kind of source selector to switch between the lamps as appropriate.

Dispersion Devices

Certain devices cause different wavelengths of light to be dispersed at different angles. When combined with an appropriate slit, they can be used to select a particular wavelength (or more precisely a narrow waveband) of light from a continuous source. Two types of dispersion device are commonly used.

The prism is known by everyone from school physics lessons for its ability to generate a rainbow from sunlight. This same principle is used in spectrophotometers. The advantages of prisms are that they are simple and inexpensive to make, but they suffer the disadvantage that the dispersion is angularly nonlinear and they are temperature sensitive.

Most modern spectrophotometers use holographic gratings. These are made from glass blanks onto which are ruled very narrow grooves. The dimensions of the grooves are of the same order as the wavelength of light, which is to be dispersed. The prepared blank is then coated with a very thin layer of aluminum to create a mirror. Light falling on the grating is reflected at different angles depending upon the wavelength. The main advantages of holographic gratings are that they give a linear angular dispersion with wavelength, they are also temperature insensitive. The disadvantage is the fact that light is reflected in different orders, which overlap as shown below. Filters must be used to ensure that only the light from one chosen reflection order reaches the detector.

A concave holographic grating combines the two functions of dispersing and focusing light at the same time.

A monochromator consists of an entrance slit, a dispersion device, and exit slit. Ideally, the output from a monochromator is monochromatic light. In practice, the output is always a band with, ideally, a symmetrical shape. The width of the band at half its height is the instrumental bandwidth (IBW).

Detectors

The detector is a device, which converts a light signal into an electrical signal. Ideally, it should give a linear response, with low noise and high sensitivity. Two types of device are commonly used in spectrophotometers.

The photomultiplier tube combines signal conversion with several stages of amplification within the body of the tube. Its main advantage is high sensitivity at low light levels. However, note that in chemical spectroscopy applications, high sensitivity means low concentrations, which in turn means low absorbances, which in turn means high intensity levels. A spectrophotometer detector must have low noise characteristic at high intensity levels, so that it is possible to determine accurately small differences between blank and sample measurements. The spectral sensitivity is determined by the nature of the cathode material. A single photomultiplier will give good sensitivity over the whole UV-visible range.

Increasingly, photodiodes are used as detectors. They have the advantage of greater dynamic range, and being solid state devices, they are more robust. In a photodiode, light falling on the semiconductor material allows electrons to flow through it, thereby depleting the charge in a capacitor connected across it. The amount of charge needed to recharge the capacitor at regular intervals is proportional to the intensity of the light. Earlier photodiodes had the disadvantage of low sensitivity in the low UV range, but modern devices have solved this problem. The limits of detection are about 150–1,100 nm for silicon-based devices.

Diode arrays are assemblies of individual photodiodes in a linear array. Self-scanned arrays have the read-out electronics included on the chip with the array. When read out, all elements of the array must be read out in series. Such arrays can have 1,024 or more elements. Externally-scanned arrays have connections to each individual element, so selected elements can be read out as required. Such arrays have limited numbers of elements; 64 is typically the maximum.

The diode arrays used in Hewlett-Packard spectrophotometers have been specifically designed for spectrophotometric applications and offer superior performance to commercially available arrays. No other spectrophotometer manufacturers manufacture their own arrays.

Optics

Optical components used to relay and focus light through the instrument are either lenses or concave mirrors.

Simple lenses are inexpensive, but suffer from chromatic aberration, that is, light of different wavelength is not focused at exactly the same point in space.

Achromatic lenses combine multiple lenses of different glasses with different refractive indices into a compound lens, which is largely free of chromatic aberration. Such lenses are used in cameras. They offer good performance, but at relatively high cost.

Concave mirrors are less expensive to make than achromatic lenses and are free from chromatic aberration. Their one disadvantage is that the aluminum surface may be easily corroded, causing a loss in efficiency.

It is worth noting that at each optical surface, 5–10% of the light is lost by absorbance or reflection.

The Conventional Scanning Spectrophotometer

The conventional scanning spectrophotometer uses a single detector and a "forward" optics configuration, that is, the dispersion device comes before the sample. A feature of such an optical system is that the sample area must be completely covered to prevent ambient light reaching the detector.

To measure at different wavelengths or to measure a spectrum with a conventional instrument with a single detector, it is necessary to rotate the dispersion device. Data acquisition, therefore, is sequential.

There are many variations in designs based on this concept but, in general, there are four important groups: Single beam, double beam, split beam, and diode array.

Single Beam

The single beam instrument is the simplest design. It is shown schematically below. To make a measurement, the blank is first placed in the instrument and measured to obtain I_o, which is usually stored within the instrument. Then, the sample is placed in the instrument (using the same cuvette) and measured to get I, and the instrument calculates the transmittance or absorbance as desired.

The advantages of the single beam design are low cost, high throughput, and hence high sensitivity, because the optical system is simple. The disadvantage is that an appreciable amount of time elapses between taking the reference (I) and making the sample measurement (I_o), so that there can be problems with drift. This was certainly true of early designs, but modern instruments have better electronics and more stable lamps, so stability with single beam instruments is now more than adequate for the vast majority of application.

Double Beam

The double beam instrument design aims to eliminate drift by measuring blank and sample virtually simultaneously. It is shown schematically below. A "chopper" alternately transmits and reflects the light beam, so that it travels down the blank and the sample optical paths to a single detector. The chopper causes the light beam to switch paths at about 50 Hz causing the detector to see a "sawtooth" signal of I_o and I, which are processed in the electronics to give either transmittance or absorbance as output.

To measure a spectrum with a double beam instrument, the two cuvettes, both containing solvent, are place in the sample and reference positions and a "balance" measurement is made. This is the difference between the two optical paths and is subtracted from all subsequent measurement. The sample is then placed in the sample cuvette and the spectrum is measured. I and I_o are measured virtually simultaneously as described above.

The advantage of the double beam design is high stability because reference and sample are measured virtually at the same moment in time, but the disadvantages are higher cost, lower sensitivity because throughput of light is poorer because of the more complex optics and lower reliability because of the greater complexity.

Split Beam

The split beam spectrophotometer is similar to the double beam spectrophotometer, but it uses a beam splitter instead of a chopper to send light along the blank and sample paths simultaneously to two separate but identical detectors. Thus, blank and sample measurements can be made at the same moment in time. Spectra are measured in the same way as with a double beam spectrophotometer.

The advantage of this design is good stability, though not as good as a double beam instrument because two detectors can drift independently, and good noise, though not as good as a single beam instrument because the light is split so that less than 100% passes through the sample.

The Diode-array Spectrophotometer

The diode-array spectrophotometer uses an array of detectors and a "reverse" optics configuration, that is, the dispersion device comes after the sample. It is shown schematically below. An advantage of the reversed optics configuration is that only light traveling along the axis from source to inlet slit before the dispersion device can reach the detector; light from other angles cannot. Thus, it is not susceptible to interference from ambient light and the sample area can be left open, making the instrument easier to use.

Light of all wavelengths falls on the diode array and is measured simultaneously, that is, data acquisition is done in parallel. A spectrum is obtained by electronically scanning the array. In principle, diode-array spectrophotometers can be designed as single-, double-, or dual-beam but, in practice, the advantages of the single beam design combine well with diode-array detection.

MAKING MEASUREMENTS

Reference or Blank Measurements

To determine a transmittance or absorbance value, it is necessary to measure I, the incident intensity, the reference measurement, and I_o the transmitted intensity of light after absorbance by the sample has occurred. If neither the cell containing the sample nor the solvent in which the sample is dissolved absorbed, it would simply require that a measurement is made first without the cuvette in the sample area (for I) and then with the cuvette in the sample area (for I_o). In practice, however, cells and solvents do absorb, so it is usual to measure the reference with the cuvette containing pure solvent in the sample area.

With a single beam instrument, always use the same cuvette for both reference and sample measurements and make regular blank measurements.

Proper Use of Cells

When making measurements, it is important to handle cuvettes carefully. For best results, adopt the following rules:

- Before starting to make measurements, check the optical surfaces of the cuvette for cleanliness and, if necessary, wipe clean with optical tissue.
- Avoid touching the optical surfaces with fingers and avoid spilling of solvent or sample on the optical surfaces.
- Do not wipe the optical surfaces clean between reference and sample measurements.
- Always place the cuvette in the sample holder with the same orientation, that is, always keep the same faces toward the source and detectors.
- Make sure that the cuvette is properly locked in place in the cell holder for each measurement. The cell holder is designed so that, when locked, the cuvette is repositioned very accurately.

QUANTITATIVE ANALYSIS

Bouguer's Law

If 100 photons of light enter a cell and only 50 emerge from the other side, the transmittance is 0.5 or 50%. If these 50 photons then pass through an identical cell, only 25 will emerge and so on.

The first mathematical formulation of this effect is generally credited to Lambert (1760), although it now appears that Bouguer first stated it in 1729. The mathematical expression is:

$$T = I/I_o = e^{-K'b}$$

where I_o is the incident intensity, I is the transmitted intensity, e is the base of natural logarithms, K' is the absorption or "extinction" coefficient (characteristic of the sample), and b is the path length (usually in cm).

Beer's Law

Beer's law is exactly analogous to Bouguer's law, except that it is stated in terms of concentration. The amount of light absorbed is proportional to the number of absorbing molecules through which the light passes. The result of plotting transmittance against path length is shown below.

Combining the two laws gives the Beer-Bouguer-Lambert law as follows:

$$T = I/I_o = e^{-K''Cb}$$

where C is the concentration of the absorbing species (usually in g/L or mg/L). This can be transformed into a linear expression by taking the logarithm.

The extinction coefficient, K'', is a characteristic of a given substance under a precisely defined set of conditions of wavelength, solvent, temperature, and other parameters. The measured extinction coefficient is also, to some extent, dependent upon the instrumental characteristics. For these reasons, predetermined values for the extinction coefficient are not usually used for quantitative analysis. In practice, a calibration or "working" curve for the substance to be analyzed is usually constructed using one or more standard solutions with known concentrations of the analyte.

To do this, a blank is required, that is, the cuvette containing the solvent used to prepare the standards and samples. The absorbance of the standards relative to the blank is then measured and the absorbance plotted against concentration. In principle, since Beer's law is linear, only one standard is required, but it is good practice to use two or more concentrations of standards, which bracket the expected sample concentrations. This enables detection of possible deviation from linearity due to instrumental or chemical effects. Using linear regression, the calibration curve is constructed from the standard measurements. Samples are then measured and their concentrations determined using the calibration curve.

Calibration Curves

Under certain circumstances, samples may not obey Beer's law and it is useful to have alternative types of calibration curve. Most spectrophotometers provide the calibration curve types shown below.

Such alternative calibration curve types should be used only with extreme caution, because they are no longer "models" of the physical process which is occurring. They are convenient approximations only.

Sources of Error

UV-visible measurements are very precise; standard deviations of 0.01% are easily achievable. However, there are many potential sources of error in quantitative analysis. They can be grouped under sample-related and instrument-related.

Sample-related sources of error include errors in preparation of standards or samples, suspended particulates that cause scattering, dirty cuvettes, etc. These errors can be minimized only by good laboratory practice.

Instrument-related errors arise from noise, drift, photometric accuracy, errors due to stray light or inadequate resolution, wavelength accuracy, and wavelength reproducibility. It is important here to distinguish between random error and bias. Random error is not reproducible whereas bias is an error that is reproducible. Instrumental errors can be classified as follows:

Random error	Bias
Noise	Insufficient resolution
Wavelength reproducibility	Wavelength accuracy
Drift	Stray light

For absolute measurements, all sources of error are significant. For relative measurements, such as UV-visible quantification, where a calibration is performed before sample measurements, bias errors cancel out and only the noise errors remain.

Note that diode-array spectrophotometer "weaknesses" are resolution, wavelength accuracy, and stray light, but the strengths are low noise, excellent wavelength reproducibility drift (when compared with conventional single-beam instruments). Thus, its characteristics match very well to the requirements for precise quantitative analysis.

NOTES

3.3 Estimation of Blood Glucose

> **LEARNING OBJECTIVES**
> At the end of session the students must be able to:
> ⊃ Know about precautions, preparations and indications of blood glucose sample collection
> ⊃ Know about different methods available for blood glucose estimations with their merits and demerits
> ⊃ Know about clinical classification of blood glucose (RBS/FBS/PP2BS/PG2BS) and interpret the laboratory results related with disorders of blood glucose metabolism/regulation
> ⊃ Perform and interpret blood glucose.

Competency B1.3.8: Discuss and interpret laboratory results of analytes with metabolism of carbohydrates.
Competency B1.3.10: Interpret the results of blood glucose levels and other laboratory investigations related to disorders of carbohydrate metabolism.
Competency B1.11.21: Demonstrate estimation of glucose in serum.
Competency IM.11.12: (Integration with medicine) Perform and interpret a capillary blood glucose test.
Domain: Knows/Shows.
Level: Knows how/Shows how/perform.
Core competency: Yes.

INTRODUCTION

The term blood sugar synonymously used for blood glucose. It is principal fuel for all the cells and is essential for brain and RBC. In day-to-day clinical practice, blood glucose estimation is indicated in following conditions:
- When patient presents with clinical features suggestive of hypoglycemia or hyperglycemia.
- Obese/overweight individuals.
- Patients presenting with hypertension, IHD, or other complications of DM, e.g. nephropathy, neuropathy, retinopathy, even an innocuous boil, or ulcer failing to resolve with conservative measures.
- All pregnant women (particularly, women with bad obstetric history, i.e. history of delivery of baby weight 4 kg, stillbirth or hydramnios, excessive weight gain during pregnancy, etc.).
- Preoperatively in all cases.
- To monitor antidiabetic therapy in diabetics.

The method used for blood processing can also influence blood glucose levels. Plasma glucose values are about 11% higher than those of whole blood when the hematocrit is normal. Arterial values are higher than venous values.

COLLECTION OF BLOOD SAMPLES

Blood sample is usually taken from cubital vein and collected in a vial containing sodium fluoride (NaF) and potassium oxalate, mixed in a ratio of 1:3 (4 mg of mixture/mL of blood). Both chemicals act as anticoagulant and NaF inhibits glycolysis in erythrocytes by inhibiting enzyme enolase. This prevents significant loss of glucose even for 2–3 days. Otherwise, glucose level is decreased at the rate of 10 mg%/h.

CLINICAL CLASSIFICATION OF BLOOD SUGAR

1. **Fasting blood sugar (FBS):** Blood sample is collected after overnight fasting of 8–12 hours.
2. **Postprandial blood sugar (PP2BS):** Blood is collected 2 hours after normal diet.

3. **Postprandial blood sugar (PG2BS):** Blood is collected 2 hours after 75 g glucose load.
4. **Random blood sugar (RBS):** Blood is collected anytime without prior preparation of the patient.

METHODS OF ESTIMATION

1. Alkaline $CuSO_4$ reduction methods:
   - Folin and Wu
   - Modified method of Asatoor and King
   - Nelson and Somogyi
   - Ortho-toluidine.
2. Enzymatic methods:
   - GOD-POD method
   - Hexokinase method.

1. **Alkaline copper sulfate reduction methods:** It has the following common steps:
   1st step: Precipitation of proteins
   2nd step: Reduction of cupric → cuprous ion and oxidation of $Fe^{2+} \rightarrow Fe^{3+}$
   3rd step: Measurement of color by colorimetry.
   Estimation of blood glucose by modified method of Asatoor and King:
   Principle: Reducing sugars in hot alkaline medium produce **enediols**, which reduces $Cu^{++} \rightarrow Cu^+$ ions that forms yellow CuOH which on further heating gives red Cu_2O which is proportional to the amount of reducing sugar. Arsenomolybdic acid reagent is added, so that oxidation of cuprous to cupric is coupled with reduction of acid → **molybdenum blue** that can be estimated colorimetrically at 610 nm. Modified method gives values close to **true value** of glucose (as compared to the original Folin and Wu method), which is achieved by diluting blood with isotonic $CuSO_4$ that contains Na_2SO_4 which prevents lysis of RBCs. The glucose diffuses out of cells leaving behind nonglucose reducing substances in the intact cells and are precipitated during deproteinization (by using 10% sodium tungstate), and removed by centrifugation.
   Ortho-toluidine method: It is the most widely used nonenzymatic method as it can be carried out without precipitation of proteins.
   Principle: The aldehyde group of glucose condenses with O-toluidine and aromatic amines to form an equilibrium mixture of glycosylamine and corresponding Schiff's base which upon further rearrangements and reactions produces a green-colored substance which is measured colorimetrically.
2. **Enzymatic methods:**
   i. **GOD-POD method (End-point):** Glucose oxidase (GOD) acts on glucose to produce gluconic acid and hydrogen peroxide. Hydrogen peroxide is producing nascent oxygen by peroxidase (POD). Nascent oxygen further reacts with a chromogen to produce colored product, which is estimated colorimetrically.

$$Glucose + O_2 \xrightarrow{Glucose\ Oxidase} Gluconic\ acid + H_2O_2$$

$$H_2O_2\ Phenol + 4\text{-aminoantipyrine} \xrightarrow{Peroxidase} \begin{array}{c}Red\ quinine\ dye\\(chromogen) + H_2O\end{array}$$

Glucose oxidase-peroxidase method (GOD-POD method):

Reagents	Blank	Standard	Test
Glucose working solution	3 mL	3 mL	3 mL
Standard	–	0.1 mL	–
Test	–	–	0.1 mL

Incubate for ~ 10–12 minutes at room temperature/37°C and read against blank at 520 nm.
Concentration of glucose in 100 mL of blood

$$\frac{OD\ of\ test}{OD\ of\ standard} \times \frac{Amount\ of\ standard}{Volume\ of\ sample/serum} \times 100$$

ii. **Hexokinase method (Kinetic method):**

$$Glucose + ATP \xrightarrow{Hexokinase} Glucose\text{-}6\text{-}PO_4 + ADP$$

$$Glucose\text{-}6\text{-}PO_4 + NAD^+ \xrightarrow{G6P\ Dehydrogenase} 6\text{-phosphogluconate} + NADH + H^+$$

The rate of formation of NADH is measured at 340 nm in spectrophotometer and also used in urine strips. This procedure is not routinely used as it is time-consuming.

DRY CHEMISTRY TECHNIQUE

Glucometer: This is a portable instrument for rapid measurement of blood glucose. Strips impregnated with glucose oxidase are allowed to react for a minute with a drop of blood obtained by fingertip puncture. The blood is then blotted and color developed in the strip is read with the instrument, which provides immediate result.

Precaution

Preservation of these strips must be done in airtight chamber.

Limitation

Slightly higher values are obtained by strip method.

Advantage

Convenient for regular sugar monitoring by diabetic patients at home and so commonly referred to as **"self-monitoring of blood glucose"** (SMBG) technique.

Fasting:
　　70–110 mg% (plasma)
　　60–100 mg% (whole blood).
　　[Recently, cutoff value of plasma glucose is reduced from 110 mg% to 100 mg% (ADA, 2004)].
　　Postprandial (PP)—100–140 mg%
　　Impaired fasting glucose (IFG)—110–126 mg%
　　Impaired postprandial sugar—140–200 mg%
　　IGT (impaired glucose tolerance)
　　Hypoglycemia < 60 mg%
Hyperglycemia:
　　Fasting > 110 mg%
　　Postprandial > 140 mg%.

IMPAIRED FASTING GLYCEMIA (IFG)

In this condition, fasting plasma sugar is above normal (between 110 mg% and 126 mg%), but 2-hour postglucose value is within normal limits (i.e. less than 140 mg/dL). The person needs no immediate treatment, but is at increased risk for development of diabetes or CVD later in life.

IMPAIRED GLUCOSE TOLERANCE (IGT)

It is, otherwise, called impaired glucose regulation (IGR). Here, blood sugar values are above normal levels, but below diabetic levels. In IGT, the FPG is between 110 mg/dL and 126 mg/dL and 2-hour PP value is between 140 mg/dL and 200 mg/dL. Such persons need careful follow-up because IGT progresses to **frank diabetes** at rate of 2% patients/year. Rarely, IGT may revert to normal condition when stress period is over. Normalization of diet, physical exercise, and dietary control are recommended in IGT.

DIAGNOSTIC CRITERIA FOR DIABETES

Fasting > 126 mg%
Postprandial > 200 mg%
　i.e. either fasting or postprandial or both greater than the above values on 2 consecutive occasions and moreover the PP value is > 200 mg% even at one occasion. All hyperglycemics are not diabetics. Hyperglycemia/hypoglycemia may be due to other conditions as:

Hyperglycemia	Hypoglycemia
• May be due to hyperactivity of endocrinal glands (thyroid and adrenals) • Emotional stress (due to production of epinephrine and norepinephrine) • Diffuse disease of pancreas • Effect of anesthesia • Mild exercise	• Tumor of pancreatic β-cells • Hypoactivity of endocrinal glands (e.g. hypothyroidism) • Impaired glucose absorption and starvation • Severe exercise decreases sugar level

GESTATIONAL DIABETES MELLITUS (GDM)

❖ This term is used when carbohydrate intolerance is noticed for the first time during pregnancy.
❖ A known diabetic patient who becomes pregnant is not included in this category. If the fasting value is more than 126 mg%, it is taken as GDM.
❖ Women with GD are at increased rate for subsequent development of frank diabetes and it is associated with increased risk of neonatal mortality.
❖ Maternal hyperglycemia causes the fetus to secrete more insulin, causing stimulation of fetal growth, and increased birth weight. After childbirth, the women must be reassessed.

Exercise 3.3.1: Estimation of glucose level in given serum sample.

Aim

....

Procedure

....

Calculations

....

Result

....

Interpretation

....

VIVA VOCE QUESTIONS: ESTIMATION OF SERUM GLUCOSE

1. How is a sample collected for blood sugar estimation in a laboratory? Why sodium fluoride is used?
2. What are the normal blood glucose values? What are fasting, PPS, and random blood samples?
3. How is the blood sugar level regulated? What are the causes of hyperglycemia?
4. What are the causes of hypoglycemia?
5. Which is the method of choice for glucose estimation?
6. Compare the reduction and enzymatic methods. Why are results higher with reduction methods?
7. What is the principle of hexokinase method? What is the main drawback of o-toluidine method?
8. What are the diagnostic criteria for diabetes mellitus?

NOTES

CASE STUDY: DIABETES MELLITUS

A 40-year-old woman, weighing about 80 kg, came to OPD of a hospital, complaining of weakness and lethargy for the last 2–3 months. She also noticed that she felt very thirst. In the night, she gets up 4–5 times for urination. When asked about the family history for such symptoms, she told that her mother had such symptoms for which she had been started with some treatment.

On physical examination, she appeared dehydrated with dryness of tongue and had mild rise of BP (146–96 mm Hg). There was no other finding.

Following were the result of investigation done:
- Blood sugar (fasting) 152 mg/dL
- Blood sugar (postprandial) 246 mg/dL
- HbA1c 14%
- Cholesterol 270 mg%
- Urea 36 mg%
- X-ray of chest and ECG were normal.

Q1. What is your diagnosis?
Q2. Why is the patient lethargic?
Q3. Why is there frequent urination?
Q4. Why does the patient have dry skin and sunken eyeballs?
Q5. What other investigation would you like to do?
Q6. What are the normal level of serum electrolytes?
Q7. Comment on the HbA1c value.
Q8. What should be the target value of blood glucose level to achieve a good diabetic control?

NOTES

Glucose Tolerance Test and Glycated Hemoglobin 3.4

> **LEARNING OBJECTIVES**
> At the end of session student must be able to know about:
> ○ Precautions, preparations and indications of oral glucose tolerance test
> ○ Glycated hemaglobin, its clinical application and interpretation.

Competency B1.3.8: Discuss and interpret laboratory results of analytes with metabolism of carbohydrates.
Competency B1.3.10: Interpret the results of blood glucose levels and other laboratory investigations related to disorders of carbohydrate metabolism.
Domain: Knows /Shows.
Level: Knows how/Shows how.
Core competency: Yes.

INTRODUCTION

The glucose tolerance test (GTT) is a well-standardized test, and is highly useful to diagnose diabetes mellitus in doubtful cases. This test is undertaken to evaluate the degree of tolerance by an individual to a glucose load under standard conditions.

INDICATIONS FOR ORAL GTT (OGTT)

1. Patient has symptoms suggestive of diabetes mellitus, but blood sugar value is inconclusive.
2. During pregnancy, big size baby is noticed or patient with bad obstetric history like miscarriages.
3. To rule out benign renal glycosuria.

PREPARATION OF THE PATIENT

1. Patient is instructed to take a good carbohydrate diet for 3 days prior to the test (at least 150 g carbohydrates). Further, a light diet should be taken on the evening prior to the investigation.
2. All drugs should be stopped for at least 2 days prior to the test.
3. Overnight fasting should be done (at least 10–14 hours).
4. Patient should be at rest mentally and physically during the test, and abstain from smoking.

PROCEDURE

- A fasting blood sample is taken in the morning. Corresponding urine sample is also collected.
- Patient is asked to drink glucose solution (75 g anhydrous glucose in 250–300 mL of water) or 1.75 g/kg body weight, maximum 75 g for children.
- The blood and urine samples are collected at 0.5-hourly interval for the rest 2.5 hours (six samples including fasting sample). Glucose is estimated in all blood samples. Urine samples are tested for glucose by Benedict's qualitative test [but the present WHO recommendation is to collect only fasting and 2-hour PP glucose load samples of blood and urine. This is, sometimes, called mini-GTT (total 2 samples only)].
- A graph of blood glucose value is plotted on vertical axis against time of collection on the horizontal axis.

FACTORS AFFECTING GTT

1. Starvation/high-fat diet
2. Exercise
3. Pregnancy (tolerance is decreased)
4. Illness: Stress causes decreased tolerance, patient's recovering from surgery, burns, or childbirth should be allowed a few weeks time before the test is carried out.

5. Physiologically: Tolerance decreases with age.
6. Drugs like OCPs, thiazide, diuretics, insulin, OHAs, and salicylates must be withdrawn before the test.
7. Liver diseases.

INTERPRETATION

The responses commonly seen are (see **Fig. 3.4.1**):

Normal response: Fasting blood sugar is normal. At 1 hour, it reaches the peak, but remains below the renal threshold (180 mg%). It returns to normal fasting level within 2–2.5 hours. None of the urine sample shows any evidence of glucose.

Diabetic curve: The fasting level is 110 mg% or more. The highest value is usually reached within 1–1.5 hours and crosses renal threshold. Glycosuria is often seen. The blood glucose level does not return to the fasting level at anytime within 2 and 2.5 hours.

Renal glycosuria: The curve is normal. Due to lowering of renal threshold, one or more urine samples contain glucose. Here, GFR is normal, but tubular reabsorption is lowered. Renal threshold is lowered physiologically in pregnancy, which is a harmless condition. It is also seen in renal tubular transport defects, e.g. Fanconi syndrome, where glycosuria, aminoaciduria, and phosphaturia are seen. In some cases, renal threshold may be increased when glucose will not appear in urine even though blood sugar is increased. Here, GFR is decreased with minimum or no impairment of tubular reabsorption. This is seen in old age arteriosclerosis and diabetic nephrosclerosis.

Lag/alimentary curve: FBG is normal. Due to rapid absorption, the maximum level is reached within 30 minutes, which crosses renal threshold. So, the corresponding urine samples show glycosuria. Hypoglycemic levels may be reached at the end of 2 hours. The curve is seen in hyperthyroidism and gastrectomy.

Flat curve of enhanced glucose tolerance: The fasting blood glucose is normal. Throughout the test, glucose level does not vary +20 mg% and is usually seen in hypothyroidism and malabsorption syndrome.

Extended GTT: Instead of ending at 2 hours and 30 minutes after taking glucose, half-hourly blood samples are taken for 4–5 hours. Partial gastrectomy patients and patients with islet cell tumors may have clinical attacks suggesting hypoglycemia, sometimes 2–4 hours after food. To clinch the diagnosis, extended GTT is done.

Intravenous GTT: Performed in patients who cannot tolerate total oral glucose like in patients of severe nausea, vomiting, and malabsorption syndrome. FBS is collected followed by injection of glucose (0.5 mg/kg body weight, maximum 35 g) in 100 mL sterile water intravenously. Samples are taken at 10 minutes interval for the next hour in these cases.

Glucose challenge test (GCT): It is done in pregnant women prior to GTT, if required. After the fasting, sample

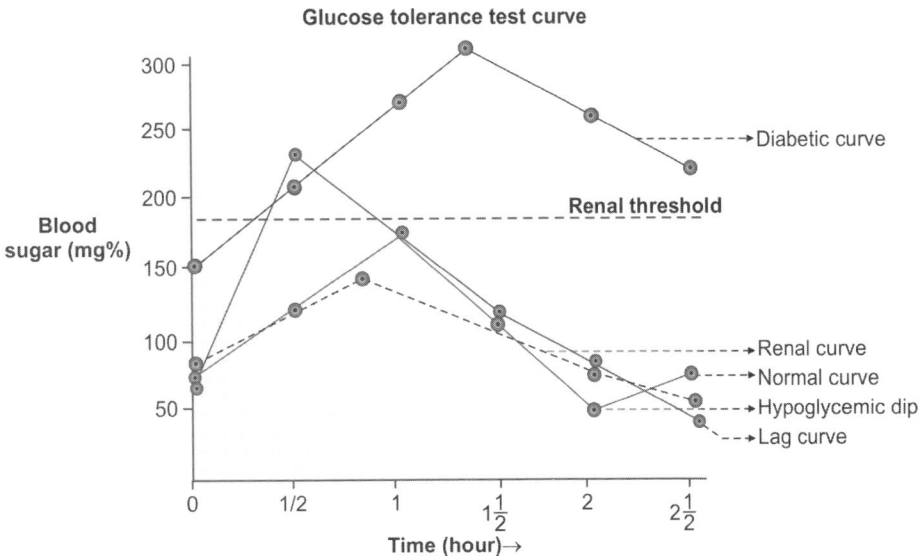

Fig. 3.4.1: Glucose tolerance test curve.

patient is given a 50-g glucose load and sample is taken at 1 hour. It should be < 140 mg%. This is only a screening test, and patients having 1-hour glucose value > 140 mg% require a full diagnostic GTT.

GLYCATED HEMOGLOBIN

- Fasting, PP, and random sugar on any day will reflect the metabolic status during that time, which is subject to alteration due to various factors including diet and medications. It does not provide an insight into the blood sugar control between two successive investigations. GHb is an indicator of integrated values for glucose over the preceding 6–8 weeks and gives information about long-term control of DM. Under physiologic conditions, HbA [which forms 90% of total Hb ($\alpha_2\beta_2$ chain)] is slowly and nonenzymatically glycated (this reaction is an exception as all glycosylation reactions in the body are enzymatic).
- The rate and extent of this reaction is dependent on average blood glucose concentration. The most abundant form of glycated Hb is HbA1c found in RBC of diabetic patients and the ratio of HbA1c to total HbA concentration is a reliable measure of degree of metabolic control in diabetic patients.

Principle

A hemolyzed preparation of whole blood is mixed continuously for 5 minutes with weak binding cation exchange resin. During this time, nonglycated Hb binds to resin. After the mixing period, a filter is used to separate the supernatant containing glycohemoglobin from the resin, which is the fast fraction, elutes first during column chromatography with **cation-exchange resin**. The % glycohemoglobin is determined by measuring the absorbance at 415 nm of glycohemoglobin (GhbA1c) fraction and total Hb (THb) fraction.

Reference Values

		Sugar values	Interpretation
i.	5–7% upto 6%	90–150 mg/dL Nondiabetic	– Good control
ii.	7–8%	150–180 mg/dL	Fair control
iii.	9–14%	180–360 mg/dL	Poor control

Interpretation

Each 1% reduction in mean HbA1c is associated with reduction in risk of deaths related to diabetes by 21%, MI by 14%, and microvascular complications by 37%.

Limitations

The interpretation of GHb depends on RBCs having a normal lifespan, i.e. 120 days. Abnormal GHb will be obtained in the following conditions:
- Hemolytic diseases } (Falsely low)
- Significant blood loss
- Iron deficiency anemia and Hb variants (falsely high).

VIVA VOCE QUESTIONS: GLUCOSE TOLERANCE TEST

1. What is the glucose tolerance test and when is it needed?
2. What are the factors which affect the GTT? What is the effect of age and diet on GTT?
3. What is impaired glucose tolerance? How is it diagnosed?
4. What is gestational diabetes mellitus? How are the pregnant females screened?
5. What do you understand by increased renal threshold? What are the causes?

NOTES

CASE STUDY: GESTATIONAL DIABETES MELLITUS

A **30-year-old woman during her second pregnancy had a glucose tolerance test and found:**

Fasting blood glucose level	1-hour glucose level	2-hour glucose level
125 mg/dL	210 mg/dL	170 mg/dL

1. Plot GTT graph with these results on the given graph sheet and comment on the GTT results.
2. What will be the results of Benedict's test with the urine sample collected along with each blood sample?
3. What is the importance of assessing the glucose tolerance in a pregnant lady?

NOTES

Liver Function Test

3.5

> **LEARNING OBJECTIVES**
>
> At the end of session student must be able to:
> ⮕ Know about different laboratory test and methods of assessment of liver functions
> ⮕ To interpret the laboratory results related with dysfunction and disorders of liver
> ⮕ Explain biochemical basis and need of lab test for liver disease.

Competency BI.6.13: Describe the functions of liver.
Competency BI.6.14: Describe the tests that are commonly done in clinical practice to assess the functions of liver.
Competency BI.11.17: Explain the basis and biochemical tests done in the following conditions: liver disease.
Competency PY.4.8: (Integration with physiology) Describe and discuss liver function test.
Domain: Knows.
Level: Knows how.
Core competency: Yes.

ASSESSMENT OF LIVER FUNCTION

Liver is vital organ of body and performs several important functions of metabolic, excretory, and detoxification nature.

Based on various function of liver, they can be grouped as:

Tests based on excretory function:
- Serum bilirubin—total and conjugated
- Urine, bile salt, and bile pigment (bilirubin and urobilinogen)
- Bromosulphalein (BSP) excretion test.

Tests based on synthetic function:
- Total serum protein
- Serum albumin and A:G ratio
- Serum cholesterol level.

Tests based on detoxification function:
- Blood ammonia
- Hippuric acid test.

Tests of liver cell damage and obstruction to bile flow:
- Aminotransferases—AST and ALT
- Alkaline phosphatase
- Gamma-glutamyl transpeptidase.

LIVER FUNCTION TEST (LFT) PROFILE

- Serum protein and A:G ratio
- Serum bilirubin—total and conjugated
- Serum SGPT (ALT) and serum SGOT (AST)
- Serum alkaline phosphatase (ALP).

LIVER FUNCTION TEST: ESTIMATION OF TOTAL PROTEIN

LEARNING OBJECTIVES

At the end of session student must be able to:
- Know about different methods available for serum protein estimations with their merits and demerits
- Interpret the laboratory results related with disorders of protein metabolism
- Explain biochemical basis and need of lab test for protein deficiency disorders.

Competency BI.11.8: Demonstrate the estimation of serum proteins.
Competency BI.11.17: Explain the basis and biochemical tests done in the following conditions: Edema.
Competency BI.11.21: Demonstrate the estimation of total protein in serum sample.
Domain: Knows/Shows.
Level: Knows how/Shows how/Perform.
Core competency: Yes.

INTRODUCTION

Plasma proteins form a complex mixture consisting of—albumin, globulin, fibrinogen, proteoglycans, lipoproteins, metalloproteins, etc. (more than 100 types of proteins are present in plasma). Fibrinogen is removed during coagulation process. Albumin is single major contributor to plasma total protein and synthesized solely in liver. Together with globulin, they constitute bulk of total proteins present in serum.

BIURET METHOD

The most commonly used method for quantitative estimation of plasma/serum proteins is Biuret method. It is simple, rapid, reliable, and reproducible method used in most automated techniques.

However, combination of total protein determination by Biuret and paper electrophoresis is most useful.

Principle: In alkaline medium, the peptide bond at least two (tripeptide) reacts with copper of biuret reagent. The cupric ions form chelates with the peptide bond of proteins (due to formation of Cu-Na-Biuret complex) forming violet or purple-colored complex, which is measured at 540 nm.

The intensity of the violet color is directly proportional to the number of peptide bonds, which in turn depends upon amount of protein in the specimen.

$$\text{Total protein} + \text{Cu} \xrightarrow{\text{Alkaline medium}} \text{Violet complex}$$

The reaction takes the name from the fact that by heating urea at high temperature, **biuret** (NH_2-CO-NH-CO-NH_2) is formed which gives similar color with alkaline $CuSO_4$ reagent.

Reagents

- **Stock biuret reagent:** Dissolve 45 g sodium potassium tartrate in about 400 mL of 2 N NaOH and add 15 g $CuSO_4$, stirring continuously. Add 5 g KI and make the volume to 1 L with NaOH. Iodide is included as an antioxidant.
- **Working biuret reagent:** Dilute 200 mL of stock reagent to 1 L with 20 mmol/L NaOH containing 5 g/L KI.
- **Standard protein solution:** Total protein standard 6 g/100 mL.

Procedure

Reagents	Blank	Standard	Test
Working biuret reagent	1.0 mL	1.0 mL	1.0 mL
Protein standard	–	10 µL	–
Sample (serum/plasma)	–	–	10 µL

Mix well and incubate for 5 minutes. Measure the absorbance of standard and sample against reagent blank at 540 nm. The final color is stable.

Calculation

Total protein concentration (g/dL)

$$= \frac{\text{OD of test}}{\text{OD of standard}} \times \frac{\text{Amount of standard}}{\text{Volume of sample used}} \times 100$$

Reference Range

Total proteins = 6–8.0 g/dL.

Precautions

- Hemolyzed serum should be avoided as it increases the color intensity.
- High sugar or lipemic in serum reduces the color intensity.

Interpretations

Hypoproteinemia

- Malnutrition (PEM) as Kwashiorkor or starvation
- Low protein intake
- Nephrotic syndrome: Loss of albumin from kidney
- Liver disease

- Celiac diseases and protein-losing enteropathies
- Chronic infection
- Untreated DM
- Hyperthyroidism and wasting disease
- Overhydration/hemodilution
- Congenital disorders like agammaglobulinemias.

Hyperproteinemia

It is usually due to increase in globulin fraction.
- Chronic infections as kala-azar
- Collagen diseases
- Bone disease as multiple myeloma
- Macroglobulinemias
- Dehydration and hemoconcentration (as diarrhea)
- Autoimmune disease (sarcoidosis and rheumatoid arthritis).

OTHER METHODS OF PROTEIN ESTIMATION

1. **Lowry method:** This is very sensitive method (100 times more sensitive than biuret method). The color development is dependent on tryptophan and tyrosine of protein molecule. The phosphotungstic-phosphomolybdic acid (Folin-Ciocalteu reagent) gives blue color with protein, but this method is not used for serum protein estimation.
2. **Dye-binding method:** Based on the ability of proteins to bind dyes as Amido Black and Coomassie Blue. The latter is commonly used for serum albumin measurement.
3. **Turbidimetric and nephelometric methods:** Used when protein level is low, viz. CSF.

Exercise 3.5.1: Estimation of serum total protein.

Aim

....

Procedure

....

Calculations

....

Result

....

Interpretation

....

LIVER FUNCTION TEST: ESTIMATION OF SERUM ALBUMIN AND A:G RATIO

LEARNING OBJECTIVES

At the end of session student must be able to:
- Know about albumin, globulin ratio and its clinical significance
- Perform and interpret serum albumin, globulin levels
- Interpret the laboratory results related with alteration in albumin, globulin ratio.

Competency BI.11.8: Demonstrate the estimation of serum albumin and A:G ratio.
Competency BI.11.17: Explain the basis and biochemical tests done in the following conditions: Nephrotic syndrome.
Competency BI.11.22: Calculate albumin:Globulin ratio.
Domain: Knows/Shows.
Level: Knows how/Perform.
Core competency: Yes.

INTRODUCTION

Serum albumin accounts for almost half of the total protein concentration in normal subjects. It acts as a carrier protein for a number of insoluble organic molecules due to hydrophobic interaction with them as: Fatty acids, urates, bilirubin, Ca^{2+}, Mg^{2+}, drugs as digoxin, antibiotics, etc.

It is the main contributor to **plasma colloidal pressure.** A number of drugs bind to albumin and absorb light at a wavelength different from the original molecule. The most specific determination of albumin is made by immunochemical techniques.

BROMOCRESOL GREEN (BCG) METHOD

It is used for the estimation of serum albumin.

Principle: This method is based on **protein error of indicators.** Binding of protein to an indicator changes its color. Albumin binds with dye **3′,3′,5′,5′-tetrabromo-m-cresol-sulfonephthalein** in acidic medium at pH 4.2. A blue-green colored complex is formed, the concentration of which is proportional to albumin present in the sample and is measured at 600 nm (600–650 nm/red filter).

$$\text{Albumin} + \text{BCG} \xrightarrow{\text{Acidic medium}} \text{Albumin-BCG complex}$$

Reagents

1. BCG reagent
2. Albumin standard—4 g/dL

Procedure

Reagents	Blank	Standard	Test
BCG working reagent	1.0 mL	1.0 mL	1.0 mL
Albumin standard	–	10 µL	–
Sample (serum/plasma)	–	–	10 µL

Mix well and incubate for 1 minute at RT and read at 600 nm.

Calculation

Serum albumin (g/dL)

$$= \frac{\text{OD of test}}{\text{OD of standard}} \times \frac{\text{Amount of standard}}{\text{Volume of sample used}} \times 100$$

Reference Range

Serum albumin = 3.7–5.3 g%.

Interpretation

Increased Serum Albumin

- Active tissue damage
- Associated with inflammatory conditions
- Dehydration/hemoconcentration
- Shock
- Malignancy.

Decreased Serum Albumin

- In malnutrition or inadequate supply of proteins
- In liver diseases (as albumin is synthesized in liver)
- In nephrotic syndrome—due to loss in urine
- In hemorrhages, excessive burns, and trauma
- Increased protein catabolism in untreated DM, hyperthyroidism, psoriasis, etc.
- Neoplastic disease and leukemia
- Plasma dilution
- Genetic defects.

Consequence of low serum albumin is **edema**. This is due to decreased plasma colloidal osmotic pressure, which favors retention of water in tissue spaces.

ESTIMATION OF SERUM GLOBULIN

Since colorimetric measurement of albumin is much simpler than that of globulin, the concentration of total protein and albumin is measured in serum, and serum globulin is obtained by difference.

Serum globulin = Total protein – serum albumin

Reference range: 2.5–3.5 g/dL.

Interpretations

Globulin is increased in:
- Acute and chronic bacterial infections
- Viral and parasitic diseases such as leishmaniasis, schistosomiasis, and malaria.
- In hepatic diseases such as:
  - Infective hepatitis
  - Cirrhosis of liver
  - Biliary cirrhosis.
- Hemochromatosis and sarcoidosis
- Plasma cell myeloma
- Lymphoproliferative diseases.

Decrease in serum globulin:
- Malnutrition
- Congenital agammaglobulinemia and acquired hypogammaglobulinemia
- Lymphatic leukemia.

A:G ratio = 1.2–2.5:1

Reversal of A:G ratio occurs in:
- Nephrotic syndrome
- Liver diseases
- In all conditions where globulin fraction is increased substantially.

Separation Technique for Protein by Zone Electrophoresis

It describes migration of charged macromolecules in porous supporting medium such as cellulose acetate strips, agarose gel, etc.

Zone electrophoresis produces electrophoretogram, a display of protein zones, each one sharply separated from neighboring zones on electrophoretic support material (medium).

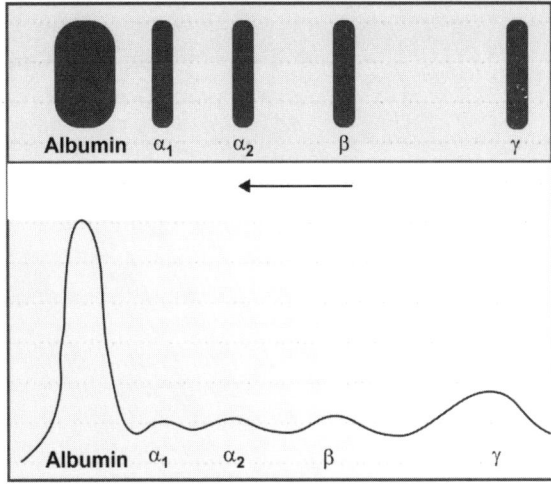

Electrophoretogram

VIVA VOCE QUESTIONS: SERUM PROTEINS AND A:G RATIO

1. What are the normal levels of total proteins, albumin, globulins, and A:G ratio?
2. Can you mention some common causes of hypoproteinemia or hyperproteinemia?
3. What is the most common clinical manifestation of hypoalbuminemia?
4. Does hemoconcentration lead to true hyperproteinemia? How can it be confirmed?
5. What is the importance of A:G ratio?
6. Name common methods of separation of different plasma protein fractions.
7. Name the different types of globulins found in plasma with few common examples of each.
8. What is microalbuminuria and its importance? Is there any proteinuria in normal individuals?

NOTES

Exercise 3.5.2: Estimation of serum albumin and A:G ratio.

Aim

....

Procedure

....

Calculations

....

Result

....

Interpretation

....

Case Study: Nephrotic Syndrome

An 8-year-old girl child attends by a pediatrician in outpatient department. She had fever and generalized edema. X-ray investigations suggested pneumonia as a possible cause of fever. She was already admitted for pneumonia four times before, and her laboratory reports were as follows:

Blood investigations	Reports	Urine investigations	Reports
Serum proteins	3 g/dL	Urine proteins	5 g/dL
Serum albumin	1.5 g/dL	Urine RBC	Absent
Fasting blood glucose	78 mg/dL	Urine pus cells	Absent
Serum cholesterol	250 mg/dL	Urine casts	Absent

Provisional diagnosis of nephritic syndrome was made, girl was treated with antibiotics and IV fluids, and later glucocorticosteroids was also given. Nutritional advice to improve protein intake was also given. Her edema improved. The specimen of urine and plasma given to you was after few weeks of treatment.

Q1. Perform the **estimation of serum albumin and A:G ratio**, write observation and inference.
Q2. Explain why there were low levels of serum protein and albumin in these patients.
Q3. Tabulated your result, correlate your results with normal values, previous laboratory reports and clinical features.
Q4. Explain biochemical basis of repeated infections and high levels of cholesterol.

NOTES

LIVER FUNCTION TEST: ESTIMATION OF SERUM BILIRUBIN

> **LEARNING OBJECTIVES**
> At the end of session student must be able to:
> - Know about serum bilirubin (total and conjugated) and its clinical significance
> - Perform and interpret serum bilirubin levels
> - Interpret the laboratory results related with alteration in bilirubin levels.

Competency Bl.11.12: Demonstrate the estimation of serum bilirubin.
Competency Bl.11.17: Explain the basis and reaction of biochemical tests done in the following condition: Jaundice.
Domain: Knows/Shows.
Level: Knows how/Perform.
Core competency: Yes.

INTRODUCTION

Bilirubin is an excretory product formed by the catabolism of heme. It is conjugated in the liver. The measurement of bilirubin in blood and urine is important in liver diseases. It is an important antioxidant.

METHOD: MALLOY-EVELYN METHOD (1937)

Principle: Bilirubin reacts with diazo reagent (diazotized sulfanilic acid) to produce colored azo pigment. At pH 5.0, the pigment is purple in color. Conjugated bilirubin, being water soluble, gives the color immediately, hence called **direct reaction.** Unconjugated bilirubin is water insoluble, hence it has to be extracted first with methanol, and then the reaction becomes positive, so called **indirect reaction.** The overall reaction is known as van den Bergh reaction.

Procedure

Reagents	Test Total	Test Direct	Standard Total	Standard Direct
Buffer	0.2 mL	0.2 mL	0.2 mL	0.2 mL
Serum	0.2 mL	0.2 mL	–	–
Standard	–	–	0.2 mL	0.2 mL
Saturated ammonium sulfate solution	0.2 mL	–	0.2 mL	–
Fresh diazo reagent	0.2 mL	0.2 mL	0.2 mL	0.2 mL
Methanol	3.2 mL	–	3.2 mL	–
Mix and allow to stand for 30 minutes in dark.				
Distilled water	–	3.4 mL	–	3.4 mL

Mix and read direct bilirubin immediately against water at 540 nm.

After 30 minutes, centrifuge the total bilirubin and read supernatant against H_2O at 540 nm.

Calculations

Concentration of total bilirubin (mg/dL):

$$= \frac{\text{OD of test}}{\text{OD of standard}} \times \frac{\text{Amount of standard}}{\text{Volume of sample}} \times 100$$

Concentration of direct bilirubin (mg/dL):

$$= \frac{\text{OD of test}}{\text{OD of standard}} \times \frac{\text{Amount of standard}}{\text{Volume of sample}} \times 100$$

Concentration of indirect bilirubin (mg/dL) = Total – Direct

Precautions

1. Avoid hemolysis as it causes false results.
2. Sample should be protected from bright light as bilirubin is photosensitive.

Interpretations

Reference range = 0.2–0.8 mg/dL.
Of this, major fraction is unconjugated which is about:
  Unconjugated → 0.2–0.6 mg/dL
  Conjugated → 0–0.2 mg/dL
  If plasma level exceeds 1 mg/dL → it is known as hyperbilirubinemia.
  Levels between 1 mg/dL and 3 mg/dL are indicative of latent jaundice (nonicteric phase).
  When levels exceed 3 mg/dL, then it diffuses into tissues producing yellowish discoloration of sclera, conjunctiva, skin, and mucous membrane producing jaundice (icteric phase).

Unconjugated Hyperbilirubinemia

i. **Physiological jaundice of the newborn or neonatal hyperbilirubinemia:** It appears 24 hours after birth. This results from an immature hepatic system for the uptake, conjugation, and secretion of bilirubin. Since the increased bilirubin is unconjugated, it is capable of penetrating the blood–brain barrier, as in newborn BBB is not developed. Phototherapy is given in these patients.
ii. **Gilbert's disease:** Defect is in the uptake of bilirubin by liver. It is an inherited defect.

iii. **Crigler-Najjar syndrome:**

Type I	Type II (Benign condition)
• Congenital nonhemolytic jaundice • Deficiency of UDP-glucuronosyltransferase enzyme • Condition is fatal and children die before the age of 2 years • Jaundice appears within 24 hours of the life • Results in a hyperbilirubinemic toxic encephalopathy or kernicterus • Unconjugated bilirubin > 20 mg/dL.	• Kernicterus does not appear • Mild defect in the bilirubin conjugating system • Bilirubin > 15 mg/dL.

iv. Hemolytic anemia
v. Hepatocellular jaundice

Conjugated Hyperbilirubinemia

Because conjugated bilirubin is water soluble, it is detectable in the urine of most patients with conjugated hyperbilirubinemia.

1. **Dubin-Johnson syndrome:**
   - Defect in the hepatic secretion of conjugated bilirubin into the bile.
   - In patients with DJ syndrome, the hepatocytes in the centrilobular area contain an abnormal pigment that has not been identified.
2. Rotor syndrome
3. Obstructive jaundice
4. Hepatocellular jaundice.

Unit 3: Quantitative Experiments and their Clinical Interpretation

Exercise 3.5.3: Estimation of serum bilirubin.

Aim

....

Procedure

....

Calculations

....

Result

....

Interpretation

....

LIVER FUNCTION TEST: ESTIMATION OF SERUM AMINOTRANSFERASES (SGOT/SGPT)

LEARNING OBJECTIVES

At the end of session student must be able to:
- Know about serum amino transaminases (SGPT/SGPT) and its clinical significance
- Perform and interpret amino transaminases (SGPT/SGPT) levels
- Interpret the laboratory results related with alteration in amino transaminases (SGPT/SGPT) levels.

Competency BI.2.2: Observe the estimation of SGOT and SGPT.
Competency BI.11.13: Demonstrate the estimation of serum aminotransaminases (SGPT/SGOT).
Domain: Knows/Shows.
Level: Knows how/Perform.
Core competency: Yes.

INTRODUCTION

- The amino transferases are a group of enzymes that catalyze the inter conversions of amino acids and α-ketoacids by transfer of amino groups.
- Aspartate amino transferase (AST) or glutamate oxaloacetate transaminase has a wide distribution, being present in heart, liver, kidneys erythrocytes and skeletal muscles.
- In cases of mild tissue damage, e.g. Liver the predominant form of serum AST is that from the cytoplasm, with a smaller amount coming from the Mitochondria e.g. viral and toxic hepatitis.
- Severe tissue damage will result in more mitochondrial enzyme being released. Elevated levels of AST can signal myocardial infarction, hepatic disease, muscular dystrophy and organ damage.
- In alcoholic patients, AST is higher than ALP since mitochondrial fraction is also released.

REITMAN AND FRANKEL'S METHOD

Principle: SGOT catalyzes transfer of amino group from L-aspartate to α-ketoglutarate with formation of oxaloacetate and glutamate. The oxaloacetate so formed is allowed to react with 2,4-DNPH to form 2,4-dinitrophenylhydrazone derivative, which is brown colored in alkaline medium. The absorbance of this hydrazone derivative is correlated to SGOT activity by plotting a calibration curve using pyruvate standard.

i. L-aspartate + α-ketoglutarate $\xrightleftharpoons{\text{SGOT}}$ Oxaloacetate + L-glutamate

ii. Oxaloacetate + 2,4-DNPH $\longrightarrow$ 2,4-dinitrophenylhydrazone (brown colored)

Reagents Provided

- Buffered substrate pH 7.4
- DNPH color reagent
- Sodium hydroxide 4 N
- Pyruvate standard (2 mM).

Precautions

- Sample should not be hemolyzed
- Sodium hydroxide 4 N is a strong corrosive reagent, handle it carefully
- Use clean and dry glassware. The presence of impurities or detergent interferes with enzyme activity.

Procedure

Pipette into two clean dry test tubes labeled blank B and test T as shown below:

Reagents	Blank	Test
Buffered substrate	0.5 mL	0.5 mL
Incubate it at 37°C for 3 minutes.		
Serum	–	0.1 mL
Mix well and incubate at 37°C for 60 minutes.		
DNPH color reagent	0.5 mL	0.5 mL
Mix well and allow to stand at RT for exactly 20 minutes.		
Distilled H_2O	0.1 mL	–
Working sodium hydroxide (dilute 1:10 with distilled water)	5.0 mL	5.0 mL

Mix well and allow to stand at RT for 10 minutes and measure the absorbance of test against blank at 505 nm.

Interpretation

Reference range = 5–35 U/mL.

The level is significantly elevated in myocardial infarction.

It is moderately elevated in liver diseases. However, a marked increase in AST may be seen in primary hepatoma.

Exercise 3.5.4: Estimation of serum aminotransferases (SGOT).

Aim

....

Procedure

....

Calculations

....

Result

....

Interpretation

....

ESTIMATION OF SGPT (ALT)

Reitman and Frankel's Method

Principle: SGPT catalyzes transfer of amino group from L-alanine to α-ketoglutarate with formation of pyruvate and glutamate. The pyruvate so formed, is allowed to react with 2,4-DNPH to form 2,4-dinitrophenylhydrazone derivative, which is brown colored in alkaline medium. The absorbance of this hydrazone derivative is correlated to SGPT activity by plotting a calibration curve using pyruvate standard.

i. L-alanine + α-ketoglutarate $\xrightleftharpoons{\text{SGOT}}$ Pyruvate + L-glutamate

ii. Pyruvate + 2,4-DNPH $\longrightarrow$ Hydrazone (brown colored)

Reagents

- Buffered substrate pH 7.4
- DNPH color reagent
- Sodium hydroxide 4 N
- Pyruvate standard (2 mM).

SGPT is unstable and hence serum should be assayed immediately or stored at –20°C.

Procedure

Pipette into two clean dry test tubes labeled blank B and test T as shown below:

Reagents	B	T
Buffered substrate	0.5 mL	0.5 mL
Incubate at 37°C for 3 minutes		
Serum	–	0.1 mL
Mix well and incubate at 37°C for 30 minutes		
DNPH color reagent	0.5 mL	0.5 mL
Mix well and allow to stand at RT for exactly 20 minutes		
Distilled H$_2$O	0.1 mL	–
Working sodium hydroxide (1:10)	5.0 mL	5.0 mL

Mix well and allow to stand at RT for 10 minutes and measure the absorbance of test against blank at 505 nm.

Interpretation

Reference range = 10–35 U/mL.

Very high values (300–1,000 U/mL) are seen in acute hepatitis, either toxic or viral in origin. Rise in ALT levels may be noticed several days before clinical signs of jaundice are manifested. Moderate increase (50–100 U/mL) of ALT may be seen in chronic liver diseases such as cirrhosis, hepatitis C, and nonalcoholic steatohepatitis (NASH).

Unit 3: Quantitative Experiments and their Clinical Interpretation

Exercise 3.5.5: Estimation of serum aminotransferases (SGPT).

Aim

....

Procedure

....

Calculations

....

Result

....

Interpretation

....

LIVER FUNCTION TEST: ESTIMATION OF SERUM ALKALINE PHOSPHATASE

LEARNING OBJECTIVES

At the end of session student must be able to:
- Know about serum alkaline phosphatase and its clinical significance
- Perform and interpret serum alkaline phosphatase
- Interpret the laboratory results related with alteration in serum alkaline phosphatase levels.

Competency BI.11.14: Demonstrate the estimation of serum alkaline phosphatase.
Domain: Knows/Shows.
Level: Knows how/Perform.
Core competency: Yes.

INTRODUCTION

- Alkaline phosphatase is present practically in all the tissues of the body. Occur at higher level in biliary intestinal epithelium, kidney tubules, bones, and placenta.
- On electrophoresis, at least 6 isoenzymes have been delineated:
  - Hepatic isoenzyme—fastest toward anode (+)
  - Bone isoenzyme
  - Placental isoenzyme
  - Intestinal isoenzyme (slowest).
  - The relative contribution of bone and liver isoenzymes to the activity of total alkaline phosphatase are markedly age dependent.
  - There is difference in level of alkaline phosphatase between different sexes at same age.

KIND AND KING METHOD

Principle: Alkaline phosphatase from serum converts phenylphosphate to phenol and inorganic phosphorus at pH 10. Now, phenol in alkaline medium reacts with 4-aminoantipyrine in presence of oxidizing agent potassium ferricyanide to form orange-red color complex measured colorimetrically.

$$\text{Phenylphosphate} \xrightarrow{\text{Alkaline PO}_4} \text{Phenol} + \text{Pi}$$

$$\text{Phenol} + \text{4-aminoantipyrine} \xrightarrow{\text{Potassium Ferricyanide}} \text{Orange-red complex}$$

Calculation

Serum alkaline phosphatase activity (KA units/dL):

$$= \frac{\text{OD of (T)} - \text{OD of (C)}*}{\text{OD of (standard)} - \text{OD of (B)}} \times \frac{\text{Concentration of standard used}}{\text{Volume of serum used}} \times 100$$

KA units correspond to the liberation of 1 mg of phenol by 100 mL of serum under optimum condition.

Interpretation

Reference range = 3–13 KA units/dL.

It is up to 2.5 times greater in children, particularly in the active growth period.

Increased ALP: Physiological conditions are:
- Children
- Growth period
- Pregnancy (3rd trimester) due to increase isoenzyme of placenta.

Pathological conditions are:
- Hepatobiliary obstruction due to stones or carcinoma
- Liver cirrhosis
- Bone disease
- Activity increase when bone regeneration is taking place. There is marked increase in rickets, osteomalacia, healing fractures, etc.
- Bone carcinoma
- Paget's disease
- Hyperparathyroidism
- Tumors of uterus
- Disease of intestinal tract.

Decrease in alkaline phosphatase occurs in:
- Achondroplasia
- Cretinism
- Scurvy
- Severe anemia
- Kwashiorkor and hypophosphatasia—inherited autosomal defect.

Unit 3: Quantitative Experiments and their Clinical Interpretation

Exercise 3.5.6: Estimation of serum alkaline phosphatase.

Aim

....

Procedure

....

Calculations

....

Result

....

Interpretation

....

CASE STUDY: JAUNDICE

Viral Hepatitis (Hepatic Jaundice)

A 14-year-old girl was admitted to the medical ward after she developed yellowish discoloration of the eye, marked loss of appetite, low-grade fever (100–101°F), nausea, and occasional vomiting in the last one week. She had pain in the right hypochondrium and urine was high colored. The stool was clay colored. She looked weak and malnourished also.

Serum LFT was done and following were the results:
- Total bilirubin—8 mg%
- Direct bilirubin—4.8 mg%
- Serum AST—980 IU/L
- Serum ALT—1,210 IU/L
- Serum ALP—20 KAU/dL

Q1. What is the type of jaundice in this patient? Explain the biochemical basis of your diagnosis.
Q2. What are the common causes of hepatic jaundice?
Q3. How the diagnosis of viral hepatitis is confirmed? Which types are associated with high morbidity and mortality?
Q4. What type of hyperbilirubinemia is seen in hepatic jaundice in viral hepatitis?
Q5. What tests would you do to detect the bile pigment in the urine sample of this patient? Explain the likely results of these tests.
Q6. What is the principle of management and what type of diet is given to these patients?
Q7. What is the usual course of viral hepatitis and its prognosis?

NOTES

Kidney Function Test 3.6

> **LEARNING OBJECTIVES**
> At the end of session student must be able to:
> - Know about different laboratory test and methods of assessment of kidney functions
> - Interpret the laboratory results related with dysfunction and disorders of kidney
> - Explain biochemical basis and need of lab test for kidney diseases.

Competency BI.6.13: Describe the functions of kidneys.
Competency BI.6.14: Describe the tests that are commonly done in clinical practice to assess the kidney functions.
Competency BI.11.17: Explain the basis and biochemical tests done in the following conditions: Proteinuria, Renal failure.
Competency PY.7.8: (Integration with physiology) Describe and discuss kidney function test.
Domain: Knows.
Level: Knows how.
Core competency: Yes.

INTRODUCTION

Kidneys perform two main functions:
A. Formation of urine as waste product
B. Production of hormones (erythropoietin, renin, and calcitriol).

Assessment of kidney function done for:
i. Extent of renal damage
ii. Monitoring the progression of renal damage
iii. Monitoring and adjusting the dose of potentially toxic drugs.

KIDNEY (RENAL) FUNCTION TEST (KFT) PROFILE

- Serum urea
- Serum creatinine
- Uric acid
- Urea clearance
- Creatinine clearance

KIDNEY FUNCTION TEST: ESTIMATION OF SERUM CREATININE AND CREATININE CLEARANCE

> **LEARNING OBJECTIVES**
>
> At the end of session student must be able to:
> - Know clinical significance serum creatinine levels
> - Perform and interpret serum creatinine levels
> - Interpret and calculate creatinine clearance.

Competency BI.11.7: Demonstrate the estimation of serum creatinine and creatinine clearance.
Competency BI.11.22: Calculate creatinine clearance.
Domain: Shows.
Level: Perform.
Core competency: Yes.

INTRODUCTION

Creatinine is the waste product formed in muscle by creatinine metabolism. Creatine is synthesized in liver by glycine, arginine, and methionine, which then passes into circulation where it is taken up by skeletal muscle for conversion to creatine phosphate, which serves as storage form of energy in skeletal muscle. Creatinine is formed spontaneously and nonenzymatically synthesized at a rate of about 2% of total body creatine per day. The amount excreted in urine remains constant for a given person and depends on his muscle bulk and is independent of protein diet. Hence, serum creatinine is a more reliable indicator of renal function.

JAFFE'S REACTION

Principle: Creatinine in alkaline medium reacts with picric acid to form orange to red-colored tautomer of **creatinine picrate**. The intensity of which is measured at 520 nm.

Procedure

	Blank	Standard	Test
Distilled water	3.2 mL	2.8 mL	2.8 mL
Serum	–	–	0.4 mL
Standard	–	0.4 mL	–
10% Na tungstate	0.4 mL	0.4 mL	0.4 mL
2/3 NH_2SO_4	0.4 mL	0.4 mL	0.4 mL
Mix and centrifuge for 5 minutes at 3,000 rpm			
Supernatant	2 mL	2 mL	2 mL
Alkaline picrate	1 mL	1 mL	1 mL

Mix well and allow to stand for 15 minutes and read at 520 nm.

Calculation

$$\text{Serum creatinine (mg\%)} = \frac{\text{OD of test}}{\text{OD of standard}} \times \frac{\text{Amount of standard used}}{\text{Volume of serum used}} \times 100$$

Disadvantages

- Lack of specificity: Only 80% of color development is due to creatinine and rest by nonspecific chromogen that reacts with picric acid like protein, ketone bodies, pyruvate, glucose, ascorbate, etc.
- It is sensitive to certain variables like pH, temperature, etc.
- Bilirubin is a negative interferrant.

Interpretation

Reference range:
- Whole blood—0.70–1.2 mg/dL
- In muscle—10 mg/dL
- Males—0.7–1.4 mg/dL
- Females—0.5–1.2 mg/dL

The range is higher in males due to more muscle mass. As ketone bodies and glucose interfere with creatinine estimation, high values may be obtained in diabetic ketoacidosis. Linearity is up to 20 mg%.

- Elevated levels are seen in renal failure and other renal diseases (similar to urea).
- No effect of age, dehydration, and catabolic states (fever, sepsis, and hemorrhage) and diet. Ratio between creatinine/urea is usually 1:20.
- Elevated levels are seen in muscular dystrophy or poliomyelitis (early stages).
- Low levels are seen in muscle wasting diseases (later stage).

CREATININE CLEARANCE

At normal level, creatinine is filtered at glomerulus, which is neither secreted nor reabsorbed by renal tubules. It does not depend on rate of flow of urine/min. The clearance of creatinine is very close to GFR (i.e. 125 mL/min).

Method

A 24-hour urine sample, blood sample, urinary creatinine, and serum creatinine are estimated.

Calculation

$$C = \frac{UV}{P}$$

U - Urinary creatinine
V - Volume of urine (mL/min)
P - Serum creatinine

Interpretation

Reference Range

- Males: 95–140 mL/min (average 120 mL/min)
- Females: 85–125 mL/min (average 110 mL/min).

Clearance value is decreased in **impaired renal function**.

Unit 3: Quantitative Experiments and their Clinical Interpretation

Exercise 3.6.1: Estimation of serum creatinine and creatinine clearance.

Aim

....

Procedure

....

Calculations

....

Result

....

Interpretation

....

KIDNEY FUNCTION TEST: ESTIMATION OF SERUM UREA AND UREA CLEARANCE

LEARNING OBJECTIVES

At the end of session student must be able to:
- Know clinical significance of serum urea levels and different methods for urea estimation
- Perform and interpret serum urea levels
- Interpret and calculate urea clearance.

Competency Bl.6.14: Describe the tests that are commonly done in clinical practice to assess the kidney functions.
Competency Bl.11.21: Demonstrate the estimation of urea in serum sample.
Domain: Know/Shows.
Level: Knows how/Shows how.
Core competency: Yes.

INTRODUCTION

Urea is the end product of protein metabolism. Protein first breaks down into amino acids, then they are deaminated in the body tissue to ammonia. The end product of fat and carbohydrate metabolism is CO_2, which is condensed with NH_3 in liver to form urea. It is mainly excreted from the body by kidney in urine. Around 15–50 g of urea is passed in urine daily.

There are various methods used for the determination of urea:

i. **Berthelot method:** This method is based on measurement of urea in serum by **urease** enzyme. Urease reacts with urea and liberates ammonia, which further reacts with phenolic chromogen and hypochlorite to form blue-green color complex (indophenol). The intensity of color is measured at 580 nm.

$$\text{Urea} + H_2O \xrightarrow{\text{Urease}} NH_3 + CO_2$$

$$NH_3 + \text{Phenol} + \text{Hypochlorite} \xrightarrow[\text{Alkaline Medium}]{\text{Nitroprusside}} \begin{array}{l}\text{Blue-green colored}\\ \text{complex}\\ \text{(Indophenol)}\end{array}$$

ii. **Diacetyl monoxime method:**
Principle: Urea reacts with diacetyl monoxime in the presence of thiosemicarbazide and cadmium ions under acidic condition to form a rose purple diazine derivative.
The intensity of the color is measured colorimetrically. This method is linear up to 300 mg%. For higher values, the blood samples should be diluted.

PROCEDURE

	Blank	Test	Standard
Urea nitrogen	5 mL	5 mL	5 mL
Distilled water	0.05 mL	–	–
Urea standard	–	–	0.05 mL
Serum	–	0.05 mL	–
DAM reagent	0.5 mL	0.5 mL	0.5 mL

Shake the test tubes and keep them in boiling water bath for 12 minutes. Cool and read it against blank at 540 nm or using green filter.

CALCULATION

Urea (mg/dL) =

$$\frac{\text{Reading of test}}{\text{Reading of standard}} \times \frac{\text{Amount of standard}}{\text{Volume of serum used}} \times 100$$

DRAWBACKS

- The reagents give strong unpleasant smell
- The reaction is complete on boiling.

COMPARISON

Although enzymatic method takes less time and less reagents, the DAM method is superior to it, as it measures urea directly, and NH_3 present in reagent and atmosphere cannot interfere, which may occur in enzymatic method.

INTERPRETATION

Reference range: 15–50 mg/100 mL.
In urine, it is 23–30 g/day.
When expressed as BUN (blood urea nitrogen), it is 7–25 mg/dL (1 mg of BUN = 2.14 mg of urea).

PHYSIOLOGICAL VARIATION

- Higher in men than women
- Varies with amount of protein in the diet
- Low in early pregnancy probably due to hemodilution (10%).

Pathological condition in which urea level is **increased** is divided in 3 groups:

1. **Prerenal:**
   - Congestive heart failure: Most important are conditions in which plasma volume/body fluid are reduced (hypovolemia)
   - Shock and hemorrhage
   - Salt and water depletion
   - Dehydration due to vomiting and diarrhea
   - Ulcerative colitis (with severe chloride ion loss)
   - Pyloric stenosis with severe vomiting
   - Intestinal obstruction
   - Burns.
2. **Renal:**
   - The blood urea can be increased in all forms of kidney disease
   - Acute glomerulonephritis
   - Nephrotic syndrome
   - Malignant nephrosclerosis
   - Chronic pyelonephritis
   - Renal tuberculosis
   - Mercurial poisoning.
3. **Postrenal:**
   Increase is due to obstruction in urine flow. This causes retention of urine and so reduces the effective filtration pressure at glomeruli.
   Enlargement of prostate
   Stone in urinary tract
   Stricture of urethra
   Tumors of bladder affecting the urinary flow.
   Decrease in urea levels:
   - Some cases of severe liver diseases
   - Cancer of liver.

UREA CLEARANCE

Clearance is defined as volume of blood or plasma, which is cleared off the substance in each minute by the kidney. It is the measure of glomerular function of kidney. Clearance is expressed as mL/min and can be calculated by using the formula:

$$C = \frac{u \times v}{p}$$

u = Concentration of substance in urine
v = Volume of urine excreted per minute
p = Concentration of substance in plasma

Clearance varies with body size and is proportional to surface area. In early stage of renal disease, clearance test offers a more sensitive index of impaired renal function than serum levels.

As a matter of fact, the plasma is not completely cleared off urea. Only about 10% of urea is removed. Consequently, 750 mL of plasma passes through the kidney/min and only 10% of urea is removed. This is equivalent to completely clearing 75 mL of plasma/min.

Maximum clearance: If the urine volume is 2 mL/min or >2 mL/min, the rate of urea eliminated is at the maximum and is directly proportional to concentration of blood urea. At this maximum urea clearance (Cm), the average normal value is 75 mL/min.

Standard clearance: When the urinary volume is less than 2 mL/min, the rate of urea eliminated is reduced because relatively more urea is reabsorbed in the tubules and is proportional to the square root of urinary volume. Such clearance is termed as standard clearance of urea (Cs). The average value is 54 mL/min.

$$C = \frac{u \times \sqrt{v}}{p}$$

Urea clearance test is expressed as % of normal maximum or of normal standard urea clearance depending on the urine output.

Method

First morning sample is discarded. Then, urine sample is collected for 24 hours using preservative thymol. Blood sample is also collected. Urea concentration is estimated in both urine and blood samples and clearance is calculated.

Interpretation

Maximum urea clearance: 60–95 mL/min.
Average: 75 mL/min.

Standard urea clearance: 40–65 mL/min.
Average: 54 mL/min.

If urea clearance is:
- 70%—normal
- 40–70%—mild impairment
- 20–40%—moderate impairment
- Below 20%—severe impairment.

Creatinine clearance test is better measure than urea clearance.
- Creatinine levels are usually constant
- It is neither secreted nor reabsorbed by kidney
- It is not dependent on the rate of flow of urine (v).

Kidney Function Test

Exercise 3.6.2: Estimation of serum urea levels and urea clearance.

Aim

....

Procedure

....

Calculations

....

Result

....

Interpretation

....

CASE STUDY: RENAL FAILURE

The following laboratory results are those of 70-year-old woman:

Laboratory investigations	Result
Serum sodium	124 mmol/L
Serum potassium	3.6 mmol/L
Blood urea	200 mg/dL
Serum creatinine	4.9 mg/dL
Serum calcium	6 mg/dL
Serum albumin	3.1 g/dL
Serum alkaline phosphatase	110 U/L
Urine albumin	++

Q1. Explain biochemical basis of patient results.

NOTES

KIDNEY FUNCTION TEST: ESTIMATION OF URIC ACID

> **LEARNING OBJECTIVES**
>
> At the end of session student must be able to:
> ○ Know clinical significance of serum uric acid levels and interpret laboratory results with alteration in uric acid levels
> ○ Explain basis and rationale of laboratory tests with alteration in uric acid levels.

Competency BI.6.14: Describe the tests that are commonly done in clinical practice to assess the kidney functions.
Competency BI.11.17: Explain the basis and rationale of biochemical test done in the following condition: Gout.
Domain: Know.
Level: Knows how.
Core competency: Yes.

INTRODUCTION

Uric acid is the catabolic end product of purines in humans. In body fluids, uric acid and monosodium urate are present. Urate is more water soluble than uric acid. Serum urate level is estimated to diagnose the inflammatory joint disease. Hyperuricemia is an important cause of gout.

Although urate is an excretory product of purines, it is an important plasma antioxidant.

ENZYMATIC METHOD: URICASE-POD (PEROXIDASE)

Principle: Uric acid is oxidized by uricase to allantoin and hydrogen peroxide, which under the influence of POD, 4-aminophenazone (4-AP) and 2,4-dichlorophenolsulfonate (2,4-DCPS) form a red quinoneimine compound.

$$\text{Uric acid} + 2H_2O + O_2 \xrightarrow{\text{Uricase}} \text{Allantoin} + CO_2 + H_2O_2$$

$$2H_2O_2 + 4\text{-AP} + \text{DCPS} \xrightarrow{\text{POD}} \text{Quinoneimine} + 4H_2O$$

The intensity of red color formed is proportional to the uric acid concentration.

Procedure

	Blank	Standard	Test
Working reagent	1.0 mL	1.0 mL	1.0 mL
Standard	–	25 µL	–
Sample	–	–	25 µL

Mix and incubate for 3 minutes at 37°C and read at 650 nm.

Read the absorbance of the sample and standard against the blank. The color is stable for at least 30 minutes.

Calculation

$$\text{mg/dL of uric acid} = \frac{\text{OD of test}}{\text{OD of standard}} \times \frac{\text{Amount of standard used}}{\text{Volume of serum}} \times 100$$

Interpretation

Reference range: 2.5–7.0 mg/dL (149–458 µmol/L)
Conversion factor = mg/dL × 59.5 = µmol/L

Interference

- Bilirubin = >170 µmol/L
- Hemoglobin = >130 mg/dL
- Ascorbic acid = >570 µmol/L

Causes of hyperuricemia:
1. **Excess production of urate:**
   A. *Primary cause (enzyme defects):*
      i. PRPP synthetase overactivity
      ii. HGPRTase deficiency
      iii. G6PD deficiency (von Gierke's disease).
   B. *Secondary cause:*
      Overproduction is due to some other disease. There is excess turnover (production/destruction) of cells leading to overproduction of uric acid.
      ♦ Cancer
      ♦ Psoriasis.
2. **Decreased excretion of uric acid (normal production):**
   ➢ Renal failure
   ➢ Lactic acidosis (alcoholism, von Gierke's disease).

Unit 3: Quantitative Experiments and their Clinical Interpretation

Exercise 3.6.3: Estimation of serum uric acid levels.

Aim

....

Procedure

....

Calculations

....

Result

....

Interpretation

....

CASE STUDY: GOUT

One executive entertained a party in which much food and alcohol had been consumed. In the next early morning, he woke up with excruciating pain in ankle. His ankle joint was swollen and red, felt hot to touch, and was very tender and stiff. The laboratory data showed: blood glucose—130 mg/dL, blood urea—38 mg/dL, serum creatinine—0.9 mg/dL, and serum uric acid—9.6 mg/dL.

Q1. What is the most probable diagnosis?
Q2. What is the management of this patient?

NOTES

Lipid Profile (Atherogenic Profile) 3.7

> **LEARNING OBJECTIVES**
>
> At the end of session student must be able to:
> - Know about different laboratory tests comes under lipid profile
> - Explain biochemical basis and need of lab test in dyslipidemia and cardiovascular disorders
> - Interpret the laboratory results related with lipid profile in various clinical context.

Competency BI.4.5/BI.4.7: Interpret laboratory results of analytes associated with metabolism of lipids.
Competency BI.11.9: Demonstrate the estimation of serum total cholesterol and HDL cholesterol.
Competency BI.11.10: Demonstrate the estimation of triglycerides.
Competency BI.11.17: Explain the basis and rationale of biochemical tests done in the dyslipidemia and myocardial infarction.
Competencies IM.2.12: (Integrated with Medicine) Choose and interpret a lipid profile and identify the desirable lipid profile in clinical context.
Domain: Knows/Shows.
Level: Knows how/Shows how/Perform.
Core competency: Yes.

LIPID PROFILE

Lipid profile is group of blood tests done to evaluate alteration in lipid levels. Total plasma lipids are estimated to be 400–600 mg/dL, of which approximately one-third is cholesterol, one-third is triacylglycerol, and one-third is phospholipids. The lipid profile (atherogenic profile) is an important blood parameter usually performed to access the possibility of coronary artery disease (CAD). It includes the estimation of:

- Serum total cholesterol ⎫
- Serum triglycerides ⎬ Estimated values
- Serum HDL-cholesterol ⎭
- Serum LDL-cholesterol ⎫ Calculated values
- Serum VLDL-cholesterol ⎭

The sample should be collected only after 12–14 hours of fasting.

LIPID PROFILE: ESTIMATION OF SERUM TOTAL CHOLESTEROL

LEARNING OBJECTIVES

At the end of session student must be able to:
- Know clinical significance of serum cholesterol levels
- Interpret laboratory results with alteration in cholesterol levels
- Perform estimation of serum cholesterol levels independently.

Competency Bl.11.9: Demonstrate the estimation of serum total cholesterol.
Domain: Shows.
Level: Perform.
Core competency: Yes.

INTRODUCTION

Serum cholesterol levels can be estimated by two methods:
a. Manual methods
b. Enzymatic methods.

Manual Methods

Liebermann-Burchard method: There are many methods, but this method is more relevant. A known amount of serum is treated with acetic anhydride and H_2SO_4 in the presence of acetic acid to form 2,4- or 3,5-cholestadienes, which combine to form their dimers or trimers which react with H_2SO_4 to form their sulfonic acid derivatives. A red color develops first, followed by blue and finally the whole solution becomes green. These colors are measured colorimetrically at 620 nm.

Enzymatic Methods

Cholesterol oxidase/peroxidase method (CHOD/POD):
Principle:

1. Cholesterol ester + H_2O $\xrightarrow{Esterase}$ Cholesterol + FFA
2. Cholesterol + O_2 $\xrightarrow{Oxidase}$ Cholestenone + H_2O_2
3. H_2O_2 + 4-aminoantipyrine + Phenol $\xrightarrow{Peroxidase}$ Quinoneimine dye (red, pink) + H_2O

The intensity of the color produced is proportional to the cholesterol concentration, which is measured colorimetrically at 520 nm or by using a green filter in spectrophotometer.

PROCEDURE

Reagent	Blank	Standard	Test
Working enzymatic reagent	3 mL	3 mL	3 mL
Standard (200 mg%)	–	0.1 mL	–
Serum	–	–	0.1 mL

Incubate all the three test tubes at 37°C for 5 minutes or 15 minutes at RT. Take the reading of the standard and test against blank at 520 nm or using a green filter.

CALCULATION

Mg of cholesterol/100 mL of serum is calculated as follows:

$$\frac{\text{Reading of test}}{\text{Reading of standard}} \times \frac{\text{Amount of standard}}{\text{Volume of serum}} \times 100$$

PRECAUTIONS

1. Test tubes must be perfectly clean and dry.
2. There must be accuracy of sample and reagent pipetting.
3. The readings must be taken after proper incubation.
4. All enzymatic reagents are stable at 2–8°C till the expiry date mentioned on the label and must not be freezed.
5. Cholesterol is stable in the specimen for 7–8 days at 2–8°C and at least for 1 month if frozen. Storage at room temperature may change the proportion of free and esterified cholesterol because of serum LCAT (lecithin-cholesterol acyltransferase), but does not affect the total cholesterol.

INTERPRETATIONS

- The normal value of total cholesterol for adults is <200 mg/dL. But Indians due to higher incidence of other risk factors as central obesity with high waist: hip ratio and Lp(a) must have even lower values of total cholesterol (i.e. <180 mg/dL).
- Values vary with age, sex, diet, and seasonal variations. Cholesterol rises with age, the rise being more marked under 50 years in men than in women (due to estrogen).
- In pregnancy, the cholesterol may reach up to 20–25% of the normal value.

PATHOLOGICAL VARIATIONS

A. **Hypercholesterolemia:**
   1. Hypothyroidism
   2. Diabetes mellitus
   3. Obstructive jaundice
   4. Nephrotic syndrome
   5. Alcohol intake
   6. Von Gierke's disease.

B. **Hypocholesterolemia:**
   - Hyperthyroidism reduces the serum cholesterol to values as low as 80–100 mg/dL.
   - Pernicious and other anemias.
   - Hemolytic jaundice and hepatocellular damage.
   - Malabsorption syndrome.
   - Wasting diseases.
   - Abeta- or hypobetalipoproteinemias.

Exercise 3.7.1: Estimation of serum cholesterol levels.

Aim

....

Procedure

....

Calculations

....

Result

....

Interpretation

....

LIPID PROFILE: ESTIMATION OF TRIGLYCERIDES

LEARNING OBJECTIVE

At the end of session student must be able to:
- Know clinical significance of serum triglyceride levels
- Interpret laboratory results with alteration in triglyceride levels
- Perform estimation of serum triglyceride levels independently

Competency Bl.11.10: Demonstrate the estimation of serum triglyceride.
Domain: Shows
Level: Perform
Core competency: Yes

INTRODUCTION

Serum triglyceride levels are done by glycerol oxidase-peroxidase method.

Principle: TG gets hydrolyzed with the help of lipase enzyme into glycerol and fatty acids.

1. Triglyceride + H_2O $\xrightarrow{\text{Lipase}}$ Glycerol + FFA
2. Glycerol + ATP $\xrightarrow{\text{Glycerol kinase}}$ Glycerol-3-PO_4 + ADP
3. Glycerol-3-PO_4 + O_2 $\xrightarrow{\text{Oxidase}}$ DHAP + H_2O_2
4. H_2O_2 + 4-aminoantipyrine + Phenol $\xrightarrow{\text{Peroxidase}}$ Quinoneimine dye (red, pink color)

The color intensity is measured colorimetrically at 520 nm.

PRECAUTIONS

i. Samples should be taken after 12–14 hours of overnight fasting.
ii. All the glasswares should be clean and properly dried.
iii. Lipemic serum should be diluted with normal saline.

PROCEDURE

Reagents	Blank	Standard	Test
Working reagent	3 mL	3 mL	3 mL
TG Standard (200 mg%)	–	0.1 mL	–
Serum sample	–	–	0.1 mL

Incubate at 37°C for 5–7 minutes or for 15 minutes at room temperature. Take the readings against blank at 520 nm.

CALCULATION

Concentration of serum triglyceride (TG) (mg/dL)

$$= \frac{\text{OD of test}}{\text{OD of standard}} \times \frac{\text{Amount of standard}}{\text{Volume of sample used}} \times 100$$

INTERPRETATIONS

Reference range = <150 mg/dL.
Increased value of TG is found in:
- Diabetes mellitus
- Nephrotic syndrome
- Hypothyroidism
- Pregnancy
- Alcoholism
- Intake of oral contraceptive pills (OCPs)
- Coronary artery disease (CAD), ischemic heart disease (IHD), and atherosclerosis.

Exercise 3.7.2: Estimation of serum triglyceride levels.

Aim

....

Procedure

....

Calculations

....

Result

....

Interpretation

....

LIPID PROFILE: ESTIMATION OF HDL-CHOLESTEROL (BY PRECIPITATION METHOD)

LEARNING OBJECTIVES

At the end of session student must be able to:
- Know clinical significance of serum HDL cholesterol levels
- Interpret laboratory results with alteration in HDL cholesterol levels
- Perform estimation of serum HDL cholesterol levels independently.

Competency Bl.11.9: Demonstrate the estimation of serum HDL-cholesterol.
Domain: Shows.
Level: Perform.
Core competency: Yes.

INTRODUCTION

Serum HDL-cholesterol levels can be estimated by precipitation method.

Principle: Phosphotungstate and $MgCl_2$ precipitate chylomicrons, VLDL, and LDL fractions. HDL fraction remains unaffected in supernatant. Cholesterol content of HDL fraction is assayed using CHOD/POD method.

Plasma or serum 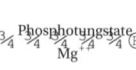 LDL + VLDL + Chylomicrons precipitated + HDL fraction in supernatant

PROCEDURE

Two step method:
1. **HDL separation:**
   Serum/plasma = 100 µL
   HDL precipitating reagent = 100 µL
   Mix thoroughly and centrifuge it. Take 50 µL of supernatant for HDL-cholesterol estimation.
2. **HDL-cholesterol estimation:**

Reagents	Test	Standard	Blank
Working cholesterol reagent	1 mL	1 mL	1 mL
HDL supernatant	0.05 mL	–	–
HDL standard (50 mg/dL)	–	0.05 mL	–

Incubate for 10 minutes at 37°C. Read at 510 nm.

CALCULATION

Concentration of serum TG (mg/dL)

$$= \frac{\text{OD of test}}{\text{OD of standard}} \times \frac{\text{Amount of standard}}{\text{Volume of sample used}} \times 100$$

INTERPRETATIONS

Reference range:
Adult male: 35–60 mg/dL
Adult female: 40–75 mg/dL
Low risk: >60 mg/dL of HDL

- HDL is reported to as "good cholesterol" or "cardioprotective" [as cholesterol present in it amounts to 30% as against 70% in LDL ("bad cholesterol")].
- HDL values < 35 mg/dL pose a high risk for development of atherosclerosis.
- For every 1 mg% drop in HDL, the risk of heart disease rises by 3%.
- HDL-C is low in Tangier disease.

LDL-C AND VLDL-C ESTIMATION (FRIEDEWALD EQUATION)

Total cholesterol = VLDL-C + LDL-C + HDL-C
1. LDL-cholesterol = Total cholesterol − [(TG/5) − (HDL)]
2. VLDL = TG/5

Reference range for LDL = <100 mg/dL.

- LDL carries about 70% of total cholesterol, transporting them to tissues, hence enhancing the risk of CHD.
- Oxidized LDL → taken up by tissue macrophages → fatty streaks → atheromatous plaques.
- Increased LDL (oxidation) is found in cigarette smokers (nicotine causes increased adipose tissue lipolysis → increased FFA and serum cholesterol), DM and insulin resistance syndrome.

Reference range of VLDL = Up to 38 mg/dL.

Desired lipid profile:
1. Cholesterol = <200 mg%
2. TG = <150 mg%
3. HDL-C = >35 mg%
4. LDL-C = <100 mg%
5. VLDL-C = Up to 38 mg%

Extended lipid profile:
ApoA1 = 94–199 mg% (rise → decreased risk of CHD)
ApoB = 60–133 mg% (rise → increased CHD risk)
Lipoprotein "a" should be <20 mg%.

Exercise 3.7.3: Estimation of serum HDL-cholesterol.

Aim

....

Procedure

....

Calculations

....

Result

....

Interpretation

....

CASE STUDY: DYSLIPIDEMIA AND MYOCARDIAL INFARCTION

A 38-year-old person was admitted in a hospital with severe chest pain. ECG was done and his blood sample was sent for the biochemical investigations. Results of CK-MB and troponin indicate that he suffered from myocardial infarction. Following investigations were done:

Laboratory investigations	Results
Total cholesterol	434 mg/dL
HDL-cholesterol	32 mg/dL
VLDL-cholesterol	39.4 mg/dL
Triglyceride	197 mg/dL
LDL-cholesterol	362.6 mg/dL
Lipoprotein(a)	38 mg/dL

Q1. Give normal range of the above serum parameter. What is lipoprotein(a) and why it is a risk factor?
Q2. How do you calculate LDL values?
Q3. How many hours of fasting are required to collect the blood sample for analysis of the above parameters?
Q4. Why HDL is known as good cholesterol and LDL is known as bad cholesterol?

NOTES

Lipid Profile (Atherogenic Profile)

Estimation of Serum Calcium and Serum Phosphorus

3.8

ESTIMATION OF SERUM CALCIUM

LEARNING OBJECTIVES

At the end of session student must be able to:
- Know clinical importance of serum calcium levels and various method for its estimation
- Perform serum calcium test independently.

Competency Bl.11.11: Demonstrate the estimation of calcium and phosphorus.
Domain: Shows.
Level: Perform.
Core competency: Yes.

INTRODUCTION

The skeleton contains 99% of body's calcium. In blood, calcium exists in 3 forms:

i. **Free or ionized (50%):** This is the physiological form. Its concentration in plasma is tightly regulated by calcium regulating hormones—PTH and vitamin D.

ii. **Protein-bound calcium (40%):** About 80% of protein-bound calcium is associated with albumin and remaining 20% to globulin (calcium binds to negative charged sites on protein, its binding is pH dependent.)

iii. **Complex (10%):** Citrate, bicarbonate, lactate, and phosphate.

Calcium is redistributed among the 3 plasma pools by alteration in concentration of protein or small anions, change in pH or change in free calcium, and total calcium in plasma. Hormones responsible for calcium homeostasis are parathormone, calcitriol, and calcitonin.

ROLE OF CALCIUM

1. In bone—combines with phosphorus to form hydroxyapatite crystals
2. Important in blood coagulation
3. Muscle contraction
4. Membrane permeability
5. Role in neuromuscular transmission
6. Needed in excitability of nerves
7. Excitability of heart.

METHOD

O-cresolphthalein complexone (OCPC) method:

Principle: Calcium ions forms a violet complex with o-cresolphthalein complexone in alkaline medium (pH 10.0). The absorbance is measured at 570 nm. The intensity of color is directly proportional to concentration of calcium.

PROCEDURE

	Blank	Standard	Test
Sample	–	–	20 µL
Calcium (standard)	–	20 µL	–
Distilled water	20 µL	–	–
Working reagent	2,500 µL	2,500 µL	2,500 µL

Mix well at room temperature for 5 minutes. Read at 570 nm (green filter) against reagent blank.

CALCULATION

$$\text{Calcium (mg\%)} = \frac{\text{OD of test}}{\text{OD of standard}} \times \frac{\text{Amount of standard}}{\text{Volume of serum}} \times 100$$

(To convert mg/dL to mmol/L, multiply with 0.25)

OTHER METHODS

The other methods are:
1. **Indirect titration:** Calcium is precipitated as oxalate and then estimated by titration with permanganate.

2. **Fluorescent method:** Based on titration of calcium fluorescent complexes.
3. **Atomic absorption spectrometry:** Sensitive and specific.
4. **Ion-selective electrode:** Measures ionized calcium.

PRECAUTIONS

- EDTA samples should not be used (only accepted anticoagulant is heparin).
- Glassware should be rinsed with 0.1 N HCl and then washed with distilled water before use.

INTERPRETATIONS

Hypocalcemia

- Osteomalacia/rickets
- Hypoparathyroidism
- Steatorrhea
- Pregnancy, lactation
- Renal cause: Nephrosis/nephritis
- Hepatocellular parenchymal disease
- Hypomagnesemia.

Hypercalcemia

- Hyperparathyroidism
- Hypervitaminosis D
- Bone neoplasm
- Milk-alkali syndrome
- Thyrotoxicosis
- Multiple myeloma
- Polycythemia vera
- Sarcoidosis.

Normal value:

Reference range in blood: 9–11 mg%.

Unit 3: Quantitative Experiments and their Clinical Interpretation

Exercise 3.8.1: Estimation of serum calcium.

Aim

....

Procedure

....

Calculations

....

Result

....

Interpretation

....

ESTIMATION OF SERUM PHOSPHORUS

LEARNING OBJECTIVES

At the end of session student must be able to:
- Know clinical importance of serum phosphorus levels
- Perform serum phosphorus test independently.

Competency B1.11.11: Demonstrate the estimation of calcium and phosphorus.
Domain: Shows.
Level: Perform.
Core competency: Yes.

INTRODUCTION

Phosphorus in serum is present in two forms:

Organic phosphorus: Phospholipids and phosphoproteins.

Inorganic phosphorus: Phosphate salts. Its level is influenced by serum calcium level. When serum PO_4^{3-} level increases, serum calcium decreases. Serum phosphate level is controlled by **parathormone**. It reduces tubular reabsorption of PO_4^{3-} in kidneys.

PRINCIPLE

Estimation by modified FISKE-SUBBAROW METHOD.

Phosphate reacts with ammonium molybdate to form ammonium phosphomolybdate, which is reduced by reducer aminonaphthol sulfonic acid (ANSA) to molybdenum blue. The blue color is measured at 710 nm.

PROCEDURE

Deproteinization of serum: 1 mL serum + 9 mL TCA (10%). Mix well and centrifuge for 10 minutes. Use supernatant. It contains all inorganic phosphates from serum.

	Blank	Standard	Test
Distilled water	5 mL	–	–
Standard PO_4^{3-} (0.04 mg/mL)	–	5 mL	–
Sample (PFF)	–	–	5 mL
Ammonium molybdate	0.4 mL	0.4 mL	0.4 mL
Perchloric acid	0.4 mL	0.4 mL	0.4 mL
ANSA	0.2 mL	0.2 mL	0.2 mL

Mix well. Keep at room temperature for 10 minutes. Read the absorbance at 710 nm or red filter.

CALCULATION

$$\frac{\text{OD of test}}{\text{OD of standard}} \times \frac{\text{Amount of standard (in mg)}}{\text{Volume of sample used (in mL)}} \times 100$$

PRECAUTIONS

1. Fasting serum to be used.
2. All the test tubes, pipettes, and other glasswares should not be contaminated with tap water, rather it should be washed with distilled water before use.
3. Blank test tube must be colorless.
4. Be careful about perchloric acid as it can cause ulceration of mucosa and damage to trachea when taken accidentally.

INTERPRETATIONS

Reference range: 2.4–4.5 mg% (1–1.5 mmol/L).

Hyperphosphatemia

1. Hypoparathyroidism
2. Renal failure
3. Hemolysis
4. Vitamin D intoxication
5. During healing of fractures.

Hypophosphatemia

1. Primary hyperparathyroidism
2. Osteomalacia/rickets
3. Ingestion of non-absorbable antacid like aluminum hydroxide
4. Dietary deficiency (Fanconi syndrome)
5. RTN (renal tubular necrosis)
6. Steatorrhea (renal tubular necrosis)
7. After injection of insulin.

Unit 3: Quantitative Experiments and their Clinical Interpretation

Exercise 3.8.1: Estimation of serum phosphorus.

Aim

....

Procedure

....

Calculations

....

Result

....

Interpretation

....

UNIT 4

Self-directed Learning Exercises

OUTLINES

4.1. pH Meter
4.2. Water Homeostasis and Estimation of Na^+ and K^+ with ISE Analyzer
4.3. Arterial Blood Gas Analyzer
4.4. Chromatography
4.5. Electrophoresis
4.6. Enzyme-linked Immunosorbent Assay
4.7. Antigen-Antibody Interaction (Immunodiffusion)
4.8. Quality Control in Clinical Laboratory
4.9. DNA Isolation from Blood and Tissue

4.1 pH Meter

> **LEARNING OBJECTIVE**
> At the end of session student must be able to know about principle, procedure and components and applications of pH meter.

Competency BI.11.16: Observe use of commonly used equipment/techniques in biochemistry laboratory including pH meter.
Domain: Knows.
Level: Knows how.
Core competency: Yes.

INTRODUCTION

- pH is defined as concentration of hydrogen ions in the solution.
- Expressed as pH = –log [H$^+$].
- Pure water has pH nearly equivalent to 7, which is considered neutral.
- Solutions having pH greater than 7 are considered basic whereas those lower than 7 are called acidic.

METHOD OF pH MEASUREMENT

- Conventionally, pH was measured with the help of litmus paper, which determined whether the solution was acidic or basic.
- Later on, pH papers were used, which gave subjective range and were not precise.
- Nowadays, pH is measured with the help of **pH meter, which** gives precise pH measurement.

pH Meter

The pH scale is a series of numbers that express the acidity as the concentration of H$^+$ ions in the solution. The normal H$^+$ ion concentration in terms of g/L is in the range of 10^{-1} to 10^{-14} (**Fig. 4.1.1**).

For simplicity, term pH = –log [H$^+$].

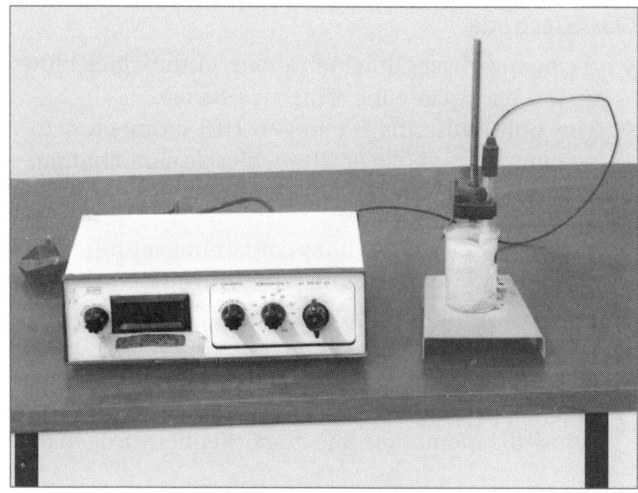

Fig. 4.1.1: pH meter.

Principle

- pH meter is based on the principle of potentiometry, i.e. measurement of electric potential generated between two electrodes of electrochemical cell due to difference in [H$^+$] concentration.
- If a metal plate is placed in a solution of its own solution, it loses ions into the solution and itself becomes negatively charged as compared to solution.

- ❖ This generates an electrical potential on the metal plate or the electrode.
- ❖ So if two different metal electrodes are connected in this way, the difference in their electrode potential can be measured as EP (electric potential).
- ❖ Hence, if one of the electrodes is a standard electrode, the electrode potential of the other can be measured by comparison.
- ❖ Electrochemical cell consists of two metal electrodes each of which is dipped into suitable but different solution connected by wire. The circuit is completed by potassium chloride/agar bridge between two liquids.
- ❖ Each metal and the electrode and the solution in which it is immersed constitute a half-cell.
- ❖ If one of the half-cells is arbitrarily assigned a fixed potential, then the potential of other half-cell may be determined relative to it. The electrode the potential of which is fixed is called standard electrode and the other electrode is called measuring electrode. The EMF (electromotive force) is defined as maximum difference in potential between two electrodes obtained when the cell current is zero.

Components of pH Meter

Glass Electrode

- ❖ It consists of very thin bulb about 0.1 mm thick blown on to a hard glass tube of high resistance.
- ❖ The bulb contains 0.1 mol/L HCl connected to a platinum wire via silver-silver chloride combination.

Calomel Electrode

- ❖ It consists of a glass tube containing saturated KCl connected to platinum wires through mercury-mercurous chloride paste.
- ❖ When the electrode is dipped in a solution, the thin glass membrane separates two solutions of two different pH values. This generates potential difference across the membrane due to transfer of hydrogen ions through glass. The EMF of electrochemical cell gives measurement of H^+ ion concentration.
- ❖ Recent instruments give measurement by digital display of pH reading.

Table 4.1.1: Commonly used buffers and their pH.

Buffers	pH values
Phosphate buffer	Around 7.0
Tris-borate-EDTA (TBE) buffer	Around 8.0
Tris-acetate-EDTA (TAE) buffer	Around 8.0
Tris-glycine (TG) buffer	More than 8.5
Lithium borate buffer	Around 8.6

Combined Electrode

- ❖ In actual instrument (pH meter), a combined electrode having both reference electrode and measuring electrode is used.
- ❖ The pH of unknown solution at 25°C is given by pH = (E − K)/0.0591, where K is constant and E is the measured potential.

Applications

- ❖ To measure pH of biological fluids.
- ❖ To adjust the pH of buffer solution, which are used in enzyme assays.
- ❖ To adjust the pH of various reagents used in biochemical assays.
- ❖ Clinical application includes measurement of pH of blood gases, CO_2, O_2, and bicarbonate using blood gas analyzer.

VIVA VOCE QUESTIONS: pH METER

1. Define pH.
2. Is pH of pure water changes with increase in temperature?
3. What do you mean by universal pH indicator?

NOTES

Water Homeostasis and Estimation of Na^+ and K^+ with ISE Analyzer

4.2

> **LEARNING OBJECTIVES**
>
> At the end of session student must be able to:
> - Know about water and electrolyte homeostasis
> - Know about principle, types and clinical interpretation of electrolyte analyzer.

Competency Bl.11.16: Observe use of commonly used equipment/techniques in biochemistry laboratory including ISE analyzer for electrolyte measurement.
Domain: Knows.
Level: Knows how.
Core competency: Yes.

INTRODUCTION

Water balance is maintained by intake and output water.

Input

- **Exogenous water:**
  - It is formed by ingested water and beverages and water content of solid foods.
  - Ingestion of water is controlled by thirst center located in hypothalamus.
- **Endogenous water:**
  - It is metabolic water produced within body.
  - 1 g each of carbohydrate, protein, and fat yield 0.6 mL, 0.4 mL, and 1.1 mL of water.

Output

- **Through urine:**
  - Major route for water loss.
  - **Regulation**
  - Water excretion through urine is controlled by ADH (antidiuretic hormone).
  - Secretion of ADH is regulated by osmotic pressure in plasma.
- **Through skin:**
  - About 450 mL/day is lost through perspiration.
  - This depends on atmospheric temperature and humidity.
  - Fever causes increased water loss through skin.
- **Through lungs:**
  Some amount of water is lost through expired air.
- **Feces:**
  About 150 mL is lost through feces.

ESTIMATION OF ELECTROLYTES (Na^+ AND K^+)

Estimation of electrolytes is done by ion-selective electrode (ISE) technique, **electrolyte analyzer principle (Fig. 4.2.1)**.

An ion-selective electrode (ISE), also known as a specific ion electrode (SIE), is a transducer (or sensor) that converts the activity of a specific ion dissolved in a solution into an electrical potential. The voltage is theoretically dependent on the logarithm of the ionic activity, according to the Nernst equation. Ion-selective electrodes are used in analytical chemistry and biochemical/biophysical research, where measurements of ionic concentration in an aqueous solution are required **(Fig. 4.2.2)**.

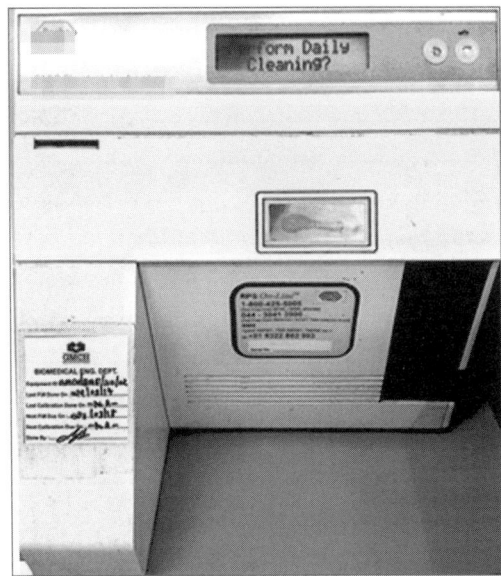

Fig. 4.2.1: ISE electrolyte analyzer.

There are four main types of ion-selective membrane used in ion-selective electrodes (ISEs)—glass, solid state, liquid based, and compound electrode.

Advantages of ISE

Ion-selective electrodes (ISE) have many advantages compared to other analytical techniques, being accurate, fast, economic, simple, and sensitive; also have an extremely wide range of application.

This technique is quiet unaffected by sample color and turbidity.

It can be used in aqueous solutions over a wide temperature range.

Nondestructive, noncontaminating, and short response time.

Limitations

Electrodes can be fouled by proteins and other organic solutes
- Interference by other ions
- Electrodes are fragile and have limited shelf-life
- Precision is rarely better than 1%.

CLINICAL INTERPRETATION OF ELECTROLYTES

Normal values:
- Serum Na^+—130–145 mEq/L
- Serum K^+—3.5–5 mEq/L

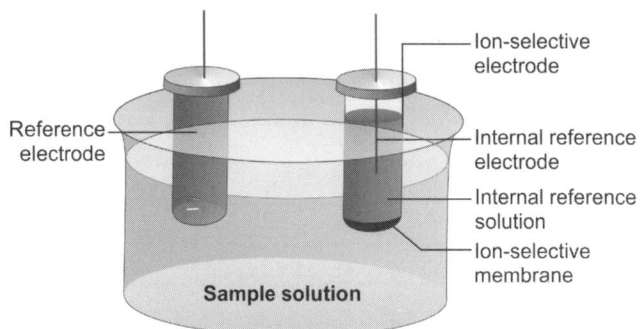

Fig. 4.2.2: Principle of ion-selective electrode.

Decreased Sodium Levels (Hyponatremia)

- Metabolic acidosis
- Salt-losing nephritis
- Addison's disease.

Increased Levels of Sodium (Hypernatremia)

- Severe dehydration
- Hyperadrenalism
- Brain injury.

Increased Potassium Levels (Hyperkalemia)

- Anorexia
- Renal tubular acidosis
- Shock
- Circulatory failure.

Decreased Levels of Potassium (Hypokalemia)

- Low dietary intake
- Loss through kidney
- Severe diarrhea or vomiting
- Increased secretion of adrenal steroids
- Diuretics.

VIVA VOCE QUESTIONS: ELECTROLYTE ANALYZER

1. Mention normal level of K^+. What are hypokalemia and hyperkalemia?
2. Mention normal level of Na^+. What are hyponatremia and hypernatremia?
3. Explain the effects of aldosterone on Na^+ and K^+ homeostasis.
4. Mention the role of antidiuretic hormone.
5. Difference between diabetes insipidus and diabetes mellitus.

NOTES

4.3 Arterial Blood Gas Analyzer

> **LEARNING OBJECTIVES**
> At the end of session student must be able to:
> ⮕ Know principle, procedure of arterial blood gas analyzer
> ⮕ Know clinical interpretation of arterial blood gas analyzer.

Competency Bl.11.16: Observe use of commonly used equipment/techniques in biochemistry laboratory including ABG analyzer.
Domain: Knows.
Level: Knows how.
Core competency: Yes.

ARTERIAL BLOOD GAS ANALYSIS

- The arterial blood gas (ABG) is a blood test that is performed using blood from an artery.
- The ABG is a test that measures the partial pressure of oxygen (pO_2), partial pressure of carbon dioxide (pCO_2), and acidity (pH) of arterial blood. In addition, arterial oxyhemoglobin saturation can also be determined.
- Such information is vital when caring of patients with critical illness or respiratory disease. As a result, the ABG is one of the most common tests performed on patients in intensive care units (ICUs) **(Fig. 4.3.1)**.

SAMPLING AND ANALYSIS

- It involves puncturing an artery with a thin needle and syringe and drawing a small volume of blood.
- The most common puncture site is the radial artery at the wrist, but sometimes the femoral artery in the groin or other sites are used. The blood can also be drawn from an arterial catheter.
- The plastic and glass syringes are used for blood gas samples. Most syringes come prepackaged and contain a small amount of **heparin** to prevent coagulation or need to be heparinized by drawing up a small amount of liquid heparin and squirting it out again to remove air bubbles.
- Once the sample is obtained, care is taken to eliminate visible gas bubbles, as these bubbles can dissolve into the sample and cause inaccurate results. The sealed syringe is taken to the laboratory for blood gas estimation.

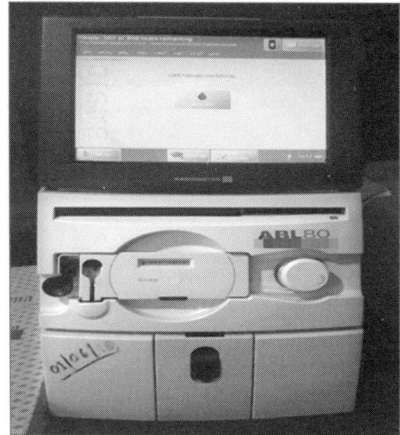

Fig. 4.3.1: ABG analyzer.

- After collection of blood sample in heparinized syringe, the transportation and analysis of the sample should be completed within 30 minutes at room temperature because body temperature can also affect arterial blood gas tensions. This is relevant in febrile or hypothermic patients, so body temperature should be recorded at the time of collection.

APPLICATION AND INTERPRETATION

Normal blood pH is between 7.34–7.42 and maintain by three tier mechanism.
- Blood buffer
- Respiratory regulation
- Renal regulation.

Though other body fluid has more or less same pH, except gastric juice **(Table 4.3.1)**.

APPLICATIONS OF ABG ANALYSIS

- To document respiratory failure and assess its severity
- To assess acid base imbalance in critical illness
- To assess response of therapeutic interventions and mechanical ventilations
- To monitor partial pressure of arterial blood gases (O_2/CO_2), base excess, and anion gap and serum bicarbonate levels in patients with acid-base imbalance **(Table 4.3.2)**.

Table 4.3.1: pH of various body fluids.

Name of body fluids	pH
CSF	7.3
Blood	7.38–7.42
Gastric juice	1.5–3.5
Lymph	Same as plasma
Seminal fluid	7.1–8.0
Bile	7.5–8.8

Table 4.3.2: ABG analytes, reference ranges, and interpretation.

Analytes	Range	Interpretation
pH (H^+)	7.34–7.44 35–45 nmol/L (nM)	• The pH indicates acidity or alkalinity of patient's blood. • If a patient is acidemic (pH < 7.35; H^+ > 45) or alkalemic (pH > 7.45; H^+ < 35)
Partial pressure of arterial oxygen (pO_2)	11–13 kPa or 75–100 mm Hg	• The state of arterial blood oxygenation is determined by the pO_2. • This reflects gas exchange in the lungs and normally the pO_2 decreases with age. This is due to decreased elastic recoil in the lung in the elderly, thereby yielding a greater ventilation-perfusion mismatch. A low pO_2 indicates that patient's blood is not properly oxygenating and is hypoxemic. This can result from hypoventilation or a mismatch of ventilation and perfusion. At a pO_2 of less than 60 mm Hg, supplemental oxygen should be administered. • At a pO_2 of less than 26 mm Hg, the patient is at risk of death and must be oxygenated immediately.
Partial pressure of arterial carbon dioxide (pCO_2)	4.7–6.0 kPa or 35–45 mm Hg	• The partial pressure of carbon dioxide (pCO_2) is an indicator of CO_2 production and its elimination during normal or constant metabolic rate. • The pCO_2 is determined entirely by its elimination through alveolar ventilation. A low pH with a high pCO_2 suggests respiratory acidosis, alternatively hypercapnia due to underventilation (or, more rarely, a hypermetabolic disorder). A low pH with a low pCO_2 suggests respiratory alkalosis, alternatively hypocapnia due to hyper- or overventilation—stimulation of respiratory center in brain.
HCO_3^-	22–26 mEq/L	• Bicarbonate is a weak base that is regulated by the kidneys as part of acid-base homeostasis. The HCO_3^- measured in arterial blood reflects the metabolic component of arterial blood. Together, CO_2 and HCO_3^- act as respiratory and metabolic buffers, respectively. $$\overset{\text{Respiratory}}{H_2O + CO_2} \overset{\text{CA}}{\leftrightarrow} H_2CO_3 \overset{\text{Metabolic}}{\leftrightarrow} HCO_3^- + H^+$$ • The HCO_3^- ion indicates a metabolic problem is present as in ketoacidosis. • A low HCO_3^- indicates metabolic acidosis and high HCO_3^- indicates metabolic alkalosis. As this value when given with blood gas results is often calculated by the blood gas analyzer.
Base excess	−2 to +2 mmol/L	• The metabolic component of the acid-base balance is reflected in the base excess. • This is a calculated value derived from blood pH and pCO_2. • It is defined as the amount of acid required to restore a liter of blood to its normal pH at a pCO_2 of 40 mm Hg. • The base excess increases in metabolic alkalosis and decreases in metabolic acidosis.
Anion gap	8 and 16 mmol/L	• The anion gap is an artificial concept that may indicate the cause of a metabolic acidosis. It represents the disparity (difference) between the major measured plasma cations (sodium and potassium) and the anions (chloride and bicarbonate). • While calculating the anion gap, potassium is usually omitted from the calculation thus: Gap = Na^+ − (Cl^- + HCO_3^-) • A raised anion gap indicates an increased concentration of lactate, ketones, or renal acids and is seen in starvation and uremia. • A normal anion gap is seen if a metabolic acidosis is due to diarrhea or urinary loss of bicarbonate. • Anion gap increases with dehydration and decreases with hypoalbuminemia.

VIVA VOCE QUESTIONS: ABG ANALYZER

1. Describe normal biological reference range of various blood gas parameters like pCO_2, pO_2, pH, HCO_3^-, etc.
2. Mention various disorders due to acid-base disturbance with examples.
3. What is anion gap and its importance?
4. Why maintenance of pH of blood and body fluids is a vital need?

NOTES

4.4 Chromatography

> **LEARNING OBJECTIVE**
> At the end of session student must be able to know about principle, procedure, types and applications of chromatography in biochemistry.

Competency BI.11.16: Observe use of commonly used equipment/techniques in biochemistry laboratory including chromatography.
Domain: Knows.
Level: Knows how.
Core competency: Yes.

INTRODUCTION

It is an analytical technique dealing with separation of closely related compounds from a mixture and includes proteins, peptides, amino acids, lipids, carbohydrates, vitamins, and drugs.

Principle and Classification

- Chromatography (Greek: *chroma*—color, *graphein*—to write) usually consists of a **mobile** and **stationary** phase.
- The mobile phase refers to mixture of substance (to be separated) dissolved in liquid or gas.
- The stationary phase is a porous solid matrix through which sample contained in mobile phase percolates.
- The interaction between mobile and stationary phase results in separation of compounds from the mixture. These interactions include the physicochemical principle as adsorption, partition, ion-exchange, molecular sieving, and affinity.
- The interaction between stationary phase and mobile phase is often employed in the classification of chromatography.
- Further, the classification of chromatography is also based either on the nature of stationary phase (paper, thin layer, and column) or on nature of both mobile and stationary phases (gas liquid chromatography).

Chromatography

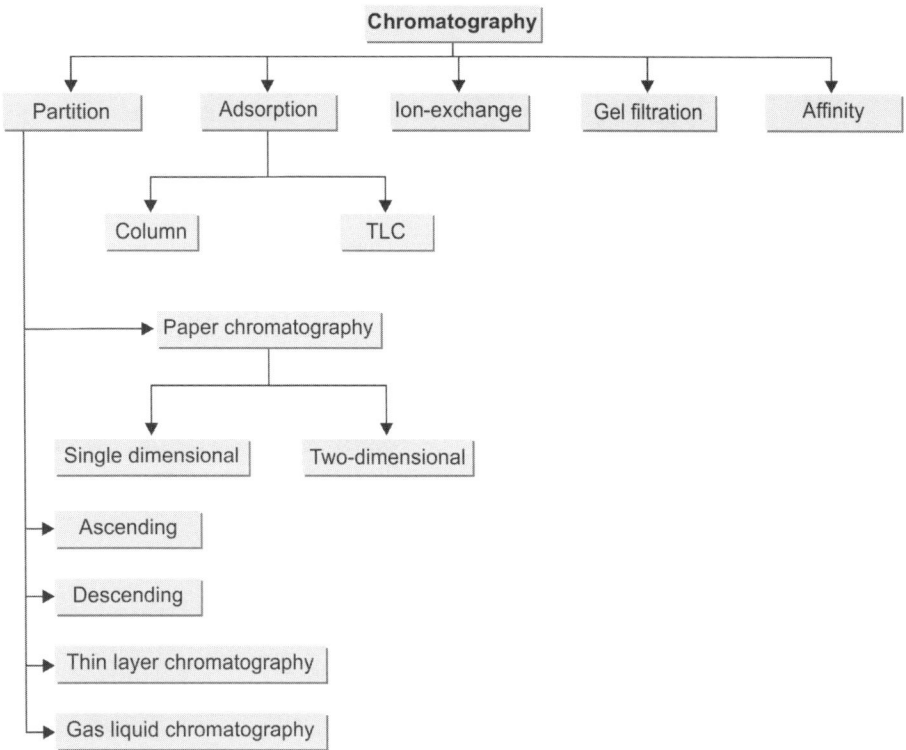

Partition Chromatography

The molecules of a mixture get partitioned between stationary and mobile phase depending on their relative affinity to each one of the phases.

Paper Chromatography

- This technique is commonly used for separation of amino acids, sugars, sugar derivative, and peptides. Amino acid in a solution can be separated by this technique and analyzed (**Fig. 4.4.5**).
- It is based on the principle of partition of compounds to be separated between two solvent phases.
- A few drops of solution containing a mixture of the compounds to be separated is applied (spotted) at one end (usually ~ 2 cm) above a strip of filter paper (Whatman No. 1 or 3). The paper is dried and dipped into a solvent mixture consisting of butanol, acetic acid, and water in **4:1:5** ratio (for separation of amino acid).
- In the **ascending** type, the solvent flows past the spot of application by capillary action. Water held back by the paper is called the **stationary** phase. Organic solvent is called **mobile phase.** In **descending** chromatography, the solvent moves downward (**Fig. 4.4.1**).
- Amino acids are soluble more in organic phase that moves faster. After a chromatographic run in closed chamber for several hours, the paper is taken out, dried, and sprayed with ninhydrin solution.
- Several purple spots develop, each representing one amino acid (ninhydrin forms purple complex with α-amino acid).
- Aromatic and branched-chain amino acids move fastest, while acidic and basic amino acids show least mobilities.
- The chemical nature of individual spots can be identified by running known standards with the unknown mixture. The migration of a substance is frequently expressed as R_f value (ratio of fronts).

$$R_f = \frac{\text{Distance travelled by the substance}}{\text{Distance travelled by solvent front}}$$

Thin Layer Chromatography (TLC)

- The principle of TLC is same as described for paper chromatography (partition). In place of paper, a thin layer of an adsorbent as silica gel, alumina, or cellulose is spread evenly on a glass plate and bound to it, serving as a support medium. The sample containing (amino acid or lipids) is applied as a small spot close to lower end of the plate, which is positioned in a close chamber containing a solvent system (benzene:acetone mixture).
- The solvent ascends through the adsorbent due to capillary action carrying the solutes to different distances.

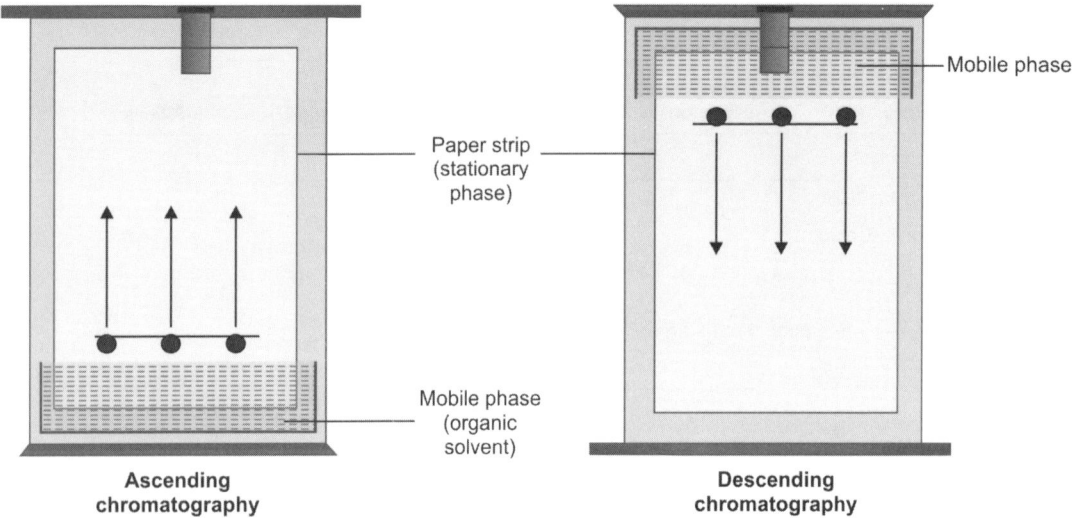

Fig. 4.4.1: Ascending and descending chromatography.

- The relative mobilities of compounds are determined by their adsorption to the matrix and also by their partition between the mobile solvent and the solvent held by the adsorbent. The compounds can be identified as distinct spots by spraying suitable chemical agents.

  The chromatographic separation is relatively rapid in TLC, which is a valuable tool for separation of lipids.

Gas-liquid Chromatography

- It is used for separation of **volatile** substances like methyl esters of fatty acid.
- In this method, a glass or metal column is filled with an inert solid-like silica gel coated with a nonvolatile liquid (e.g. silicone oil). The column is maintained at 170–225°C.
- The vaporized sample is introduced at one end of the column and is carried through the column by a stream of inert gas like nitrogen. Each component in the sample moves through the column, determined by its partition coefficient between mobile phase and the stationary liquid. Individual components of the sample emerging out of the column in gas phase are detected by physical or chemical means. The chromatographic pattern will consist of number of peaks, each corresponding to one compound. The area under each peak is proportional to concentration of the compound. Time taken to emerge out of the column is characteristic of each compound **(Fig. 4.4.2)**.

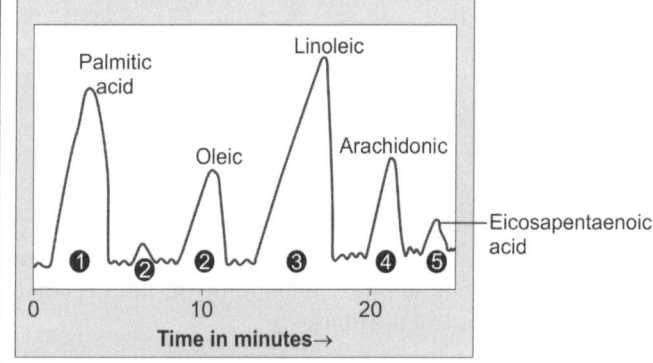

Methyl esters of separation of fatty acid found esterified to cholesterol in rat plasma by gas-liquid chromatography

Fig. 4.4.2: Gas-liquid chromatography.

Adsorption Column Chromatography

- The adsorbents as silica gel, alumina, charcoal powder, and calcium hydroxyapatite are packed into a column in a glass tube. This serves as stationary phase. The sample mixture in a solvent is loaded on this column. The individual components get differentially adsorbed on to the adsorbent. The elution is carried out by a buffer system (mobile phase). The individual components come out of the column at different rates, which may be separately collected and identified.
- For instance, amino acids can be identified by ninhydrin colorimetric method **(Fig. 4.4.3)**.

Ion-exchange Chromatography

- A better method of separation of amino acids is ion-exchange chromatography in which commercially available ion-exchange resins are used.
- Ion-exchange resins consist of a support matrix to which cationic and anionic groups are covalently attached.

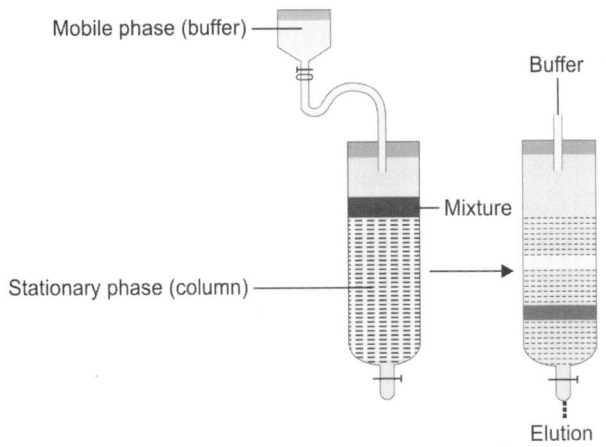

Fig. 4.4.3: Adsorption column chromatography.

- The ion exchangers carrying fixed negative charges are **cation exchangers** and those with positive charges are **anion exchangers.**
- In amino acid separation, strong ion-exchange resins based on a matrix of polystyrene cross-linked with divinylbenzene are used.
- Ion-exchange resin column is prepared by filling a glass tube with Na⁺ form of cation-exchange resin. A solution of amino acids of pH 3.0 is allowed to seep into top layer of the column. The extent of binding of amino acid is a function of net charge on the molecules. At pH 2.0, all amino acids carry net positive charge and are bound. The bound amino acids are eluted from the column by buffers of increasing pH.
- Individual amino acids emerge at different stages, which can be collected as fractions. The order of elution of amino acid is as follows:
  - Aspartic acid, threonine, serine, glutamic acid, and proline at pH 3.25.
  - Glycine, alanine, cysteine, valine, methionine, isoleucine, leucine, tyrosine, and phenylalanine at pH 4.25.
  - Lysine, histidine, and arginine at pH 5.28.
- Individual amino acid in the fractions collected can be estimated by ninhydrin/dansyl chloride reaction.

Gel Filtration Chromatography/Molecular Exclusion Chromatography

- Another physical method known as **gel permeation** or **filtration** technique is used in determination of: (i) size of proteins, and (ii) also for separation of proteins based on molecular weight.
- When proteins in solutions are passed through a column packed with cross-linked hydrophilic polymer of acrylamide, dextran, or agarose, they emerge in the effluent in order of decreasing size. By varying the degree of cross-linking, the range of size in which separation takes place can be fixed anywhere in the region of 10^3–10^7 Daltons.
- When a molecule of large size is allowed to flow through the column bed, it does not penetrate into the polymer matrix and thus moves faster and will have a short path length through the column. On the other hand, a small molecule permeates through the polymer gel and travels a longer path before it comes out.

  V_t = Total volume, occupied by whole column
  V_o = Void volume → volume excluding the volume occupied by the polymer gel.
  V_e = Elution volume of a particular small molecule.
- The elution position of molecules differing in size is given by following equation:

$$K = \frac{V_e - V_o}{V_t - V_o}$$

- A plot of V_e/V_o against log molecular weights will give a straight line, which can be used for determination of the molecular weight of unknown protein once V_e is determined. This is done by using a calibrated column with substances of known molecular weight (**Fig. 4.4.4**).

Affinity Chromatography

- It is a powerful technique used in isolation of proteins, enzymes, vitamins, nucleic acid, drugs, antibodies, and hormone receptors.
- In this procedure, a specific ligand which has high affinity for a particular protein is covalently attached to an insoluble matrix, like cross-linked agarose.
- The mixture containing the protein to be isolated is allowed to permeate through the column containing the affinity matrix. All other proteins come out of the column in washing.
- The protein is then desorbed by eluting with a substrate that weakens the interaction between ligand and the protein. For example, many dehydrogenases are specifically bound to the dye Cibacron Blue at pH 7.0

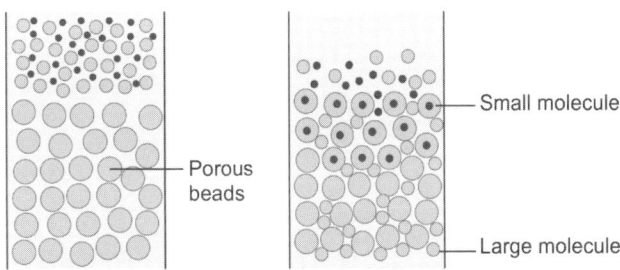

Fig. 4.4.4: The principle of gel filtration chromatography.

when it is immobilized. The bound enzymes can be eluted with substrates, NADH or NADPH.

❖ IgG is specifically bound through its Fc region to a protein known as protein A at pH 8.0. Lowering the pH of eluting medium results in release of IgG.

Fig. 4.4.5: Chromatography equipment.

High Performance Liquid Chromatography (HPLC)

❖ In general, chromatographic techniques are slow and time-consuming. The separation can be greatly improved by applying high pressure in range of 5,000–10000 psi (pounds/square inch), which reduces time to few minutes. Hence, this technique is also referred to as high pressure liquid chromatography. It is a refinement based on the same principles of regular column chromatography.

VIVA VOCE QUESTIONS: CHROMATOGRAPHY

1. What is chromatography and what is the principle of chromatographic process?
2. What is the meaning of term Rf (retention factor) in chromatography and on which factors this value depends?
3. What is the advantage of chromatography over other technique?

NOTES

4.5 Electrophoresis

> **LEARNING OBJECTIVE**
> At the end of session student must be able to know about principle, procedure, components and applications of electrophoresis in biochemistry.

Competency Bl.11.16: Observe use of commonly used equipment/techniques in biochemistry laboratory including electrophoresis.
Domain: Knows.
Level: Knows how.
Core competency: Yes.

ELECTROPHORESIS

The amount of charged particles (ions) in an electric field resulting in their migration toward oppositely charged electrode is known as **electrophoresis.** It is a widely used analytical technique for separation of biological molecules as plasma proteins, lipoproteins, and immunoglobulins.

Different Types of Electrophoresis

I. **Zone Electrophoresis:**
   a. Paper
   b. Gel
- In its simplest form known as **zone electrophoresis,** filter paper, agar gel, or cellulose acetate strips are used as a support medium **(Fig. 4.5.1)**.
- A few microliters solution of a mixture of charged particles (e.g. proteins) are applied at the support medium in form of a narrow band.
   a. **Paper electrophoresis:** For separation of serum proteins, Whatman No. 1 filter paper, Veronal, or Tris buffer at pH 8.6 and stains: Amido black or Bromophenol blue are employed **(Fig. 4.5.2)**.

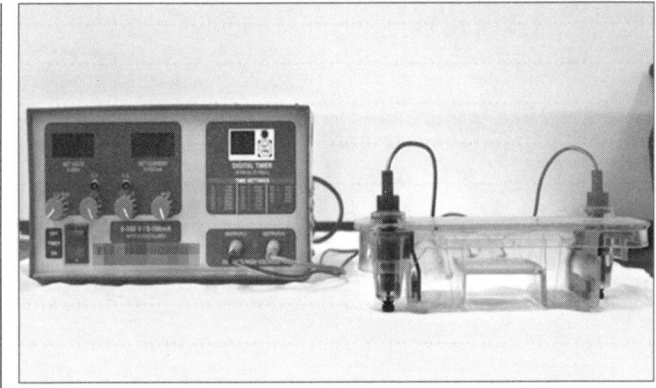

Fig. 4.5.1: Electrophoresis equipment.

   b. **Gel electrophoresis (polyacrylamide):** As support medium → polymers or acrylamide cross-linked with methylene bisacrylamide is used. Polyacrylamide, as a thin slab, a glass plate, or short columns is used. The resolution is far better by this technique.
- The medium is placed in an electrophoretic chamber containing compartments on either side of the platform on which the support medium is placed. The

Electrophoresis

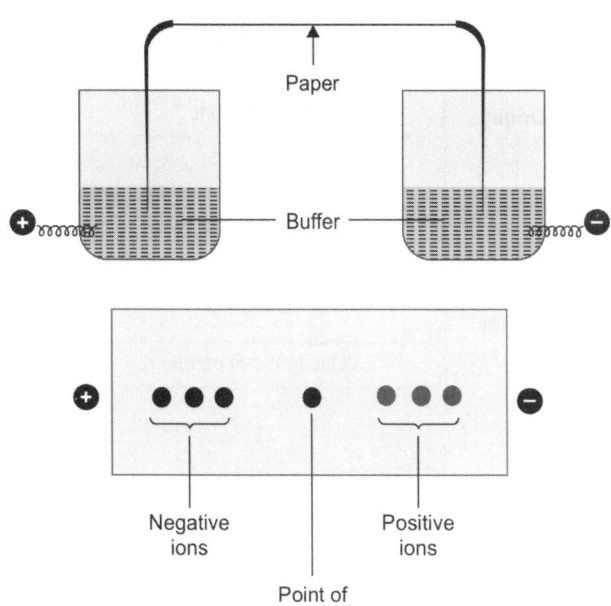

Fig. 4.5.2: Paper electrophoresis.

compartments are filled with equal volume of buffer solution of suitable pH. The electrodes dip in the buffers. A direct current is allowed to flow through. After a run, the medium is taken out.

- ❖ Depending on charge and densities, proteins will move toward cathode or anode. The protein bands are stained with dyes.
- ❖ The serum proteins by paper electrophoresis are separated into 5 distinct bands—albumin, α_1, α_2, β, and γ globulin.

1. SDS-PAGE (Determination of molecular weight):

- ❖ A much simpler method for estimation of molecular weight of proteins and protein subunits is by electrophoresis in medium of polyacrylamide gel (PAG) in presence of sodium dodecyl SO_4 (SDS) **(Fig. 4.5.3)**.
- ❖ SDS interacts with nonpolar interior of protein molecules and unfolds the native structure. The charged SO_4^{2-} groups of **protein-SDS complex** contribute negative charge. With proteins containing disulfide bridges, prior reduction with mercaptoethanol is necessary.
- ❖ On electrophoresis, protein-SDS complex moves toward anode. The rate of mobility is inversely proportional to size of the protein.
- ❖ To determine molecular weight of an unknown protein, several standard proteins of known molecular weight are subjected to electrophoresis along with unknown protein under identical conditions.
- ❖ Protein bands are stained with suitable dyes (Coomassie Brilliant Blue, Amidoschwartz, or silver stain) and mobilities are estimated. When log of molecular weight is plotted against mobility, a straight line is obtained. From such a plot for standard proteins, molecular weight of an unknown protein can be estimated.

2. Isoelectric Focusing:

- ❖ This technique is primarily based on immobilization of molecules at isoelectric pH during electrophoresis.
- ❖ Stable pH gradients are set up (usually in gel) covering the pH range to include the isoelectric points of the components in a mixture.

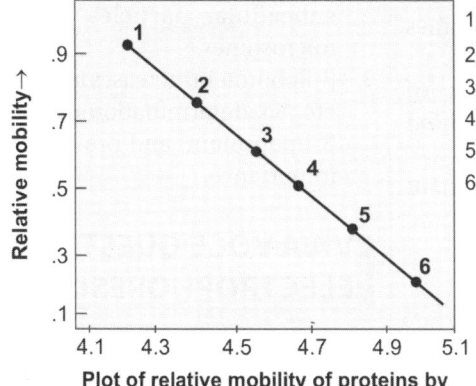

1) Lysozyme (14 kDa)
2) β-Lactoglobulin (18.4 kDa)
3) Trypsinogen (24 kDa)
4) Pepsin (35 kDa)
5) Ovalbumin (45 kDa)
6) Bovine serum albumin (68 kDa)

Plot of relative mobility of proteins by SDS-PAGE against log molecular weight

Fig. 4.5.3: Protein migration on SDS page.

- ❖ This technique is conveniently used for purification of proteins.

3. Immunoelectrophoresis:
- ❖ This technique involves combination of principles of electrophoresis and immunological reactions.
- ❖ It is useful for the analysis of complex mixtures of antigens and antibodies.
- ❖ The complex proteins of biological samples (human serum) are subjected to electrophoresis. The antibody (antihuman immune serum from rabbit or horse) is then applied in a trough parallel to electrophoretic separation. The antibodies diffuse and when they come in contact with antigens, precipitation occurs, resulting in formation of precipitin bands which can be identified.

Lipoproteins: Separation and Function

Lipoprotein consists of lipid core →
- ❖ Triacylglycerol
- ❖ Cholesterol ester

Surrounded by polar coat shell of
- ❖ Phospholipids
- ❖ Cholesterol
- ❖ Apoproteins

i. Triacylglycerol 45%
ii. Phospholipids 35%
iii. Cholesterol and cholesterol esters 15%
} Combination of all these forms ↓
iv. Free fatty acids and proteins <5% Lipoprotein complex

- ❖ Since pure fat is less dense than water, the proportion of lipid:protein in lipoproteins in plasma is determined by ultracentrifugation.
- ❖ The density of lipoproteins increases as protein content rises and lipid content falls, and size of particle becomes smaller.
- ❖ Lipoproteins may be separated on basis of electrophoretic properties and may be identified accurately by immunoelectrophoresis.
- ❖ Four major groups of lipoproteins which are important physiologically in clinical diagnosis and in some metabolic disorders of lipid metabolism are:
  i. Chylomicrons
  ii. VLDL/pre-β-lipoprotein
  iii. LDL/β-lipoprotein
  iv. HDL/α-lipoprotein
  ➤ Chylomicrons and VLDL:
    Triacylglycerol : 50%
    Cholesterol : 23%
    } Increased in atherosclerosis and coronary thrombosis

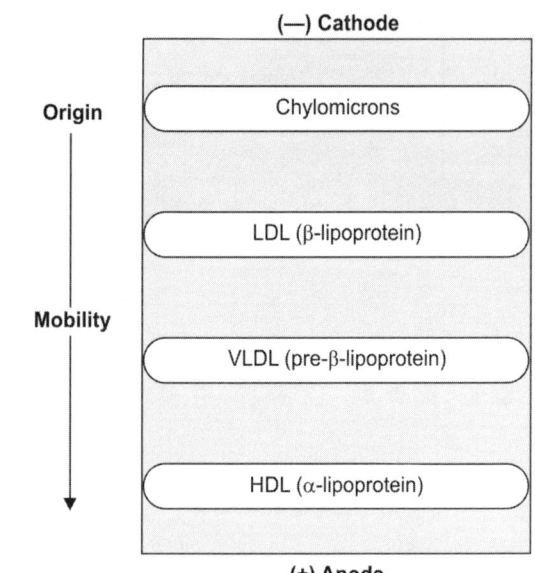

Fig. 4.5.4: Electrophoresis of plasma (serum) lipoproteins.

➤ LDL:
  Cholesterol: 46%
  Phospholipids: 23%
  } Increase in coronary thrombosis
➤ HDL:
  Phospholipids: 27%
  Proteins: 45%
- ❖ Protein moiety in lipoprotein is known as apoprotein (60% of HDL and 1% of chylomicron) **(Fig. 4.5.4)**.

Importance

1. Transport and deliver lipids to tissues.
2. Maintain structural integrity of cell surface and subcellular particles such as mitochondria and microsomes.
3. β-lipoprotein increases in → severe DM, atherosclerosis, etc. So, determination of relative concentration of α, β-lipoprotein, and pre-β-lipoprotein is of diagnostic importance.

VIVA VOCE QUESTIONS: ELECTROPHORESIS

1. What is electrophoresis and what are the important applications of electrophoresis?
2. What are the different types of electrophoresis?
3. Which type of electrophoresis is used for isoenzyme separation?

NOTES

4.6 Enzyme-linked Immunosorbent Assay

> **LEARNING OBJECTIVE**
> At the end of session student must be able to know about principle, procedure, components and applications of ELISA in biochemistry.

Competency BI.11.16: Observe use of commonly used equipment/techniques in biochemistry laboratory including ELISA.
Domain: Knows.
Level: Knows how.
Core competency: Yes.

INTRODUCTION

- This is a simplified and modified version of RIA using enzyme linked to Ab instead of radiolabeled antigen. It, therefore, eliminates the need of radiation counters. The Ag-Ab binding is detected by an enzyme catalyzed color reaction measurable by simple colorimeter or spectrophotometer.
- The technique is most frequently used in the form of solid phase ELISA. Either Ag or Ab is bound to microtiter wells when reaction is carried out by addition of a suitable substrate.
- Commonly used enzymes are—alkaline phosphatase, peroxidase, β-galactosidase, and glucose-6-phosphate dehydrogenase. A single molecule of enzyme converts 10–10^4 molecules of substrate → products/min.
- Can be used for both qualitative and quantitative measurements. A standard curve is plotted for quantitative measurements and is quite safe and cheaper than RIA.
- Several types as indirect, sandwich, and competitive ELISA are available.

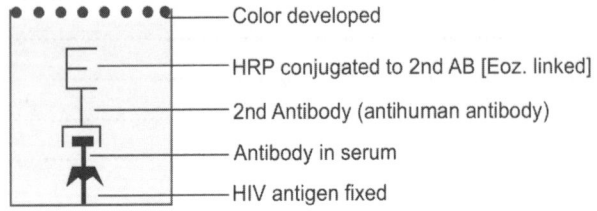

SANDWICH ELISA/ANTIGEN DETECTION

- In this technique, Ag is sandwiched between two antibodies. The microtiter plate/wells are coated with Ag-specific Abs.
- It is reacted with antigen first and then a second Ab specific for a different epitope of antigen is added. The 2nd Ab is enzyme tagged and so leads to colored product on addition of substrate. This is then measured by an ELISA reader.
  For example, suppose we want to quantitate AFP (α-fetoprotein-antigen seen in liver cancer) in serum of a patient.
- Specific Ab is fixed to the well of a microtiter plate. The patient's serum is added in the well and incubated for 30 minutes at 37°C. By this time, if serum contains Ag, it is fixed on Ab.
- Excess antigen and other proteins are washed out.
- Then anti-AFP antibody tagged with HRP (horseradish peroxidase) is added. If antigen is already fixed, the antibody-HRP-conjugate will be fixed in the well. The color reagent containing H_2O_2 and diaminobenzidine (DAB) is added. The reaction is:

$$H_2O_2 \xrightarrow{HRP} H_2O + O \text{ (nascent oxygen)}$$

$$\downarrow$$

Diaminobenzidine ⟶ Oxidized DAB
(Colorless)   (brown color)

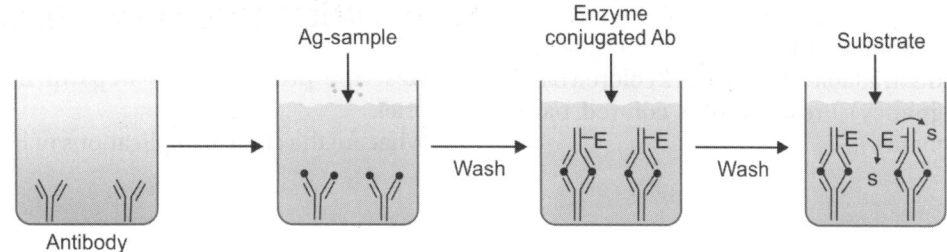

Fig. 4.6.1: Principle of sandwich ELISA.

The color developed is proportional to antigen in the serum. So, intensity of the color is measured from which concentration of antigen is calculated. Other chromogens used are:
- NBT (nitroblue tetrazolium): Blue color
- NPP (nitrophenyl phosphate): Yellow color **(Fig. 4.6.1)**.

INDIRECT ELISA: ANTIBODY DETECTION

- This is useful to detect small quantities of Abs in blood (e.g. detection of HIV antibody).
- In patients with AIDS, the HIV virus produces specific Ab. To detect HIV antibody, following method is used.
- Ag from HIV is coated in wells of a microtiter plate. Patient's serum is added and incubated. If it contains antibody, it is fixed. The wells are washed to remove excess Abs.
- Next, a second Ab (Ab against human Ig) conjugated with HRP is added. Then, color reagent containing H_2O_2 and diaminobenzidine is poured over.
- A brown color, if develops, is proportional to Ab concentration **(Fig. 4.6.2)**.

COMPETITIVE ELISA

- In this, sample containing antigen is incubated with excess but known amount of Ab **(Fig. 4.6.3)**.
- Depending upon Ag concentration, the reaction mixture will have some amount of free Ab. More antigen will, therefore, result in less free Ab and vice versa.
- When added to microtiter well coated with Ag, free Ab will bind to it.
- Enzyme-tagged specific secondary Ab is then added and detected by color development.

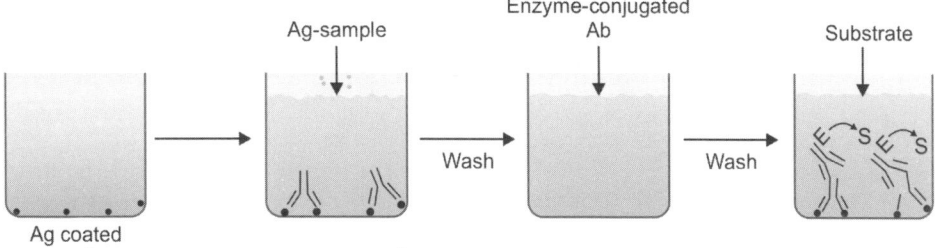

Fig. 4.6.2: Principle of indirect ELISA.

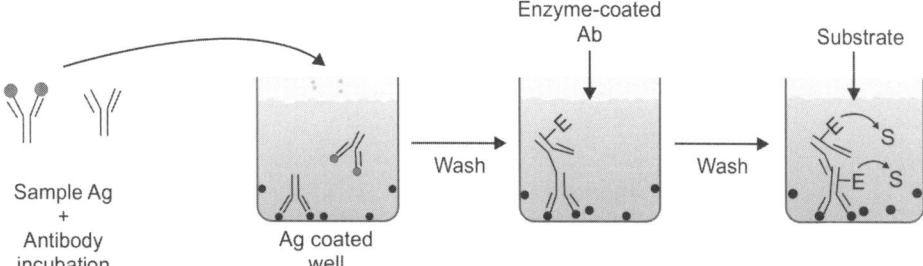

Fig. 4.6.3: Principle of competitive ELISA.

- In competitive ELISA, absorbance will be indirectly proportional to amount of antigen [this forms the basis of pregnancy kits available. Presence of 2 colored bands (positive pregnancy) Presence of 3 colored bands (negative pregnancy)].

VIVA VOCE QUESTIONS: ELISA

1. Describe principle and steps of ELISA process in brief.
2. What are the clinical applications of ELISA?

NOTES

4.7 Antigen-Antibody Interaction (Immunodiffusion)

> **LEARNING OBJECTIVE**
> At the end of session student must be able to know about principle, procedure and components and applications of immunodiffusion.

Competency BI.11.16: Observe use of commonly used equipment/techniques in biochemistry laboratory including immunodiffusion.
Domain: Knows.
Level: Knows how.
Core competency: Yes.

ANTIGEN-ANTIBODY INTERACTION (IMMUNODIFFUSION)

In vivo, B cells recognize antigen and bind it through its receptor, which is an Ab molecule. This enables the B cells to act as antigen-presenting cells and ultimately gets itself activated and becomes an effector cell, i.e. plasma cell or memory cell.

- Further soluble antibodies, present in plasma and other body secretions, neutralize and enhance the elimination of antigen by binding to them. So, antigen-antibody interaction is essential for immune mechanism *in vivo*.
- *In vitro*, this interaction forms the basis of highly specific and sensitive immunological techniques used in medical diagnosis and research.

Epitope and Paratope

The antigen and antibody on coming in contact with each other interact in small discrete areas rather than the whole molecule. The part of antibody reacting with antigen is called **paratope** and part of antigen reacting with antibody is called **epitope**. The paratope and epitope are very crucial and determine the closeness or the strength of interaction.

An antigen molecule usually has many epitopes often with different specifications. It is seen that B cells and T cells do not recognize the same epitopes on antigen molecules but different epitopes and so called (a) B-cell epitopes and (b) T-cell epitopes.

Antigen-Antibody Interaction

The antigen-antibody binding is a specific bimolecular interaction similar to enzyme-substrate interaction. This binding is of noncovalent and reversible nature. The bonds involved in this interaction are:
a. Hydrogen bonds
b. Electrostatic interactions
c. Hydrophobic interactions
d. Van der Waals interactions.

The strength of Ag-Ab binding is indicated by: (i) Affinity and (ii) Avidity.
i. The tight fit between antigen epitope and paratope determines affinity of antibody.
ii. The strength of interactions between multivalent antibody and antigen is called avidity (**Figs. 4.7.1A to C**).

Immunochemical Techniques

Agglutination and Precipitation

- Antigen reacts with specific antibodies to form antigen-antibody complex. Depending upon the size of this complex and nature of antigen, two types of reactions

Antigen-Antibody Interaction (Immunodiffusion)

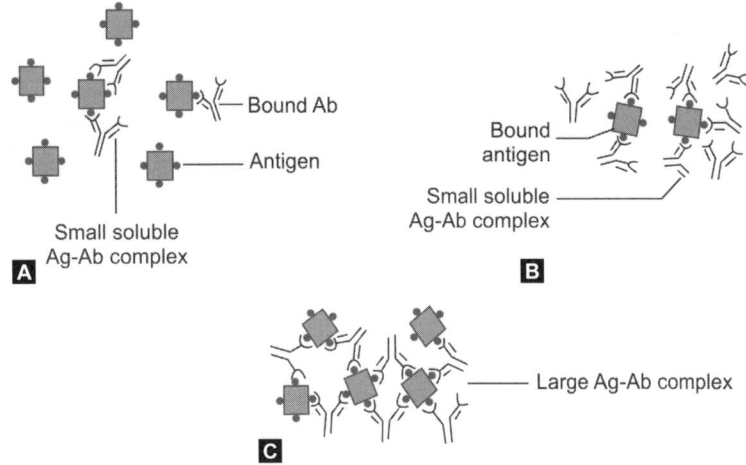

Figs. 4.7.1A to C: Effects of antigen and antibody concentration on precipitation reaction. (A) Antigen excess; (B) Antibody excess; (C) Optimum Ag-Ab concentration.

are known. They are called (1) Precipitation reaction and (2) Agglutination reaction.

- When a **soluble** antigen having at least 2 or more antibody binding sites react with corresponding antibody in suitable concentration, it leads to the formation of large antigen-antibody complex, which is visible to naked eye as a precipitate. Such reactions are called **precipitation** reactions.
- When antigen is of **particulate** nature, its reaction with the antibody leads to visible clumping called **agglutination** (e.g. routine blood group testing reaction).

Precipitins and Precipitation Reactions

- Antibodies leading to precipitation reaction are precipitins and are usually bivalent or polyvalent.
- Both antigens and antibodies have to be in optimal concentration, so that each antigen and antibody molecule can react with many molecules to form a large complex appearing as precipitate.
- In excess of either the antigen or antibody, small and soluble complexes are formed that do not appear as precipitate.
- For example, when antigen is in excess, most of it is present in free unbound form in the supernatant. Only few antigen molecules are bound to the antibody forming small and soluble antigen antibody complexes.
- Similarly in antibody excess, only few antibodies are bound to antigen and most of them are again present in free form in the supernatant.
- It is at the optimal ratio of the Ag and Ab concentration that the precipitation occurs.

The optimal ratio can be easily determined, e.g. if an increasing antigen amount is added to a series of test tubes, each containing fixed amount of Ab and each tube is centrifuged to concentrate and measure the precipitate, then tube showing maximum precipitate indicates optimal ratio of antigen and antibody. In graphical representation, this zone of optimal ratio is called **equivalence zone** (Fig. 4.7.2).

- Precipitation ratio can be used to detect precipitin Ab in body fluids and other samples. The reaction can be carried out in test tubes in liquid state or in gels.
- In liquid state, the precipitate is formed which is detectable. In gel, the precipitation reaction is visible as "lines" at the site of Ag-Ab interaction and these days, mostly gel is used for precipitation.
- Most protein antigen and some carbohydrates and carbohydrate-lipid conjugate antigens produce precipitin Ab.

Techniques based on precipitin Ab are:
- Immunodiffusion
- Radial immunodiffusion

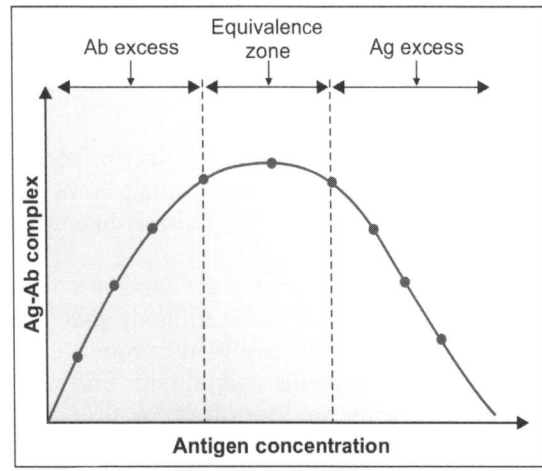

Fig. 4.7.2: Plot of Ag-Ab complex with Ag concentration.

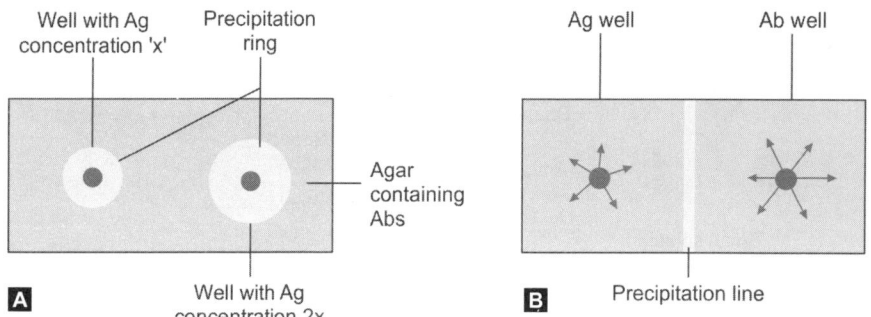

Figs. 4.7.3A and B: Immunodiffusion. (A) Radial immunodiffusion; (B) Double immunodiffusion.

- Immunoelectrophoresis
- Countercurrent immunoelectrophoresis
- Rocket immunoelectrophoresis.

Immunodiffusion:

- In this technique, antigen is allowed to diffuse into a semisolid gel containing antibody. As the antigen diffuses away from the original site, i.e. the well containing antigen, its concentration gradually decreases and it forms a precipitation ring or zone where the optimal Ag-Ab concentration ratio is achieved. The area of ring or zone is proportional to antigen concentration.
- A standard curve of the area of precipitin ring or zone obtained with known concentration of the antigen can be plotted. Subsequently, the concentration of unknown samples can be calculated from this curve.

Radial Immunodiffusion (Mancini method):

In this, antigen is placed in a well and allowed to diffuse into the semisolid agar containing Ab to produce a radial precipitation zone as a circular spot. Its area is proportional to Ag concentration **(Fig. 4.7.3A)**.

Double Immunodiffusion (Ouchterlony method):

Both Ag and Ab are allowed to diffuse gradually toward each other, thereby establishing a concentration gradient. A precipitin line is seen in equivalence zone **(Fig. 4.7.3B)**.

Immunoelectrophoresis:

- Here, the antigen mixture is first electrophoresed to separate its components. Then, a trough is cut into the gel parallel to the line of electrophoresis on one or both sides of slides.
- The specific antibody to the target antigen is placed in the trough. On diffusion of the antibody, precipitation lines are produced in the equivalence zone.
- Different serum proteins components and multiple myeloma Ig can be identified in this manner **(Fig. 4.7.4)**.

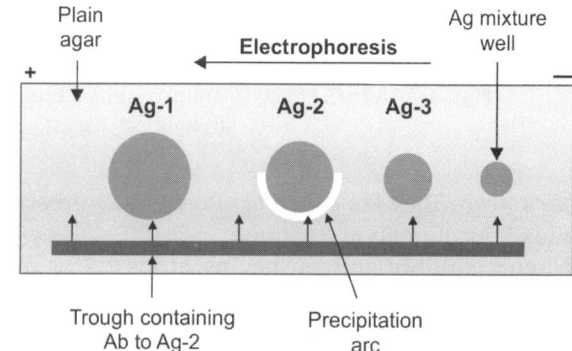

Fig. 4.7.4: Immunoelectrophoresis.

Modifications of Immunoelectrophoresis:

- Rocket
- 2D

Rocket Electrophoresis:

- The gel contains known antibody concentration. Several wells are cut on one side of plate and increasing concentrations of antigens are placed in them. Ag well side is connected to negative terminus and opposite side to the positive terminus and antigen is electrophoresed.
- Precipitation appears in form of rocket—the height being proportional to concentration of antigen. Unknown sample is simultaneously run with a series of known samples to evaluate its Ag concentration **(Fig. 4.7.5)**.

2D Electrophoresis:

The antigen is subjected to electrophoresis twice, first as such on an agar plate without any antiserum containing gel, which is then laid over the plate after 1st electrophoresis. The 2nd electrophoresis is then done in a direction at right angle to the first one. The precipitation zones are detected as overlapping peaks, their size being proportional to concentration of antigen. The number of peaks is equal to number of proteins in the mixture.

Immunoelectrophoresis requires that the proteins to be separated and identified are electrically charged

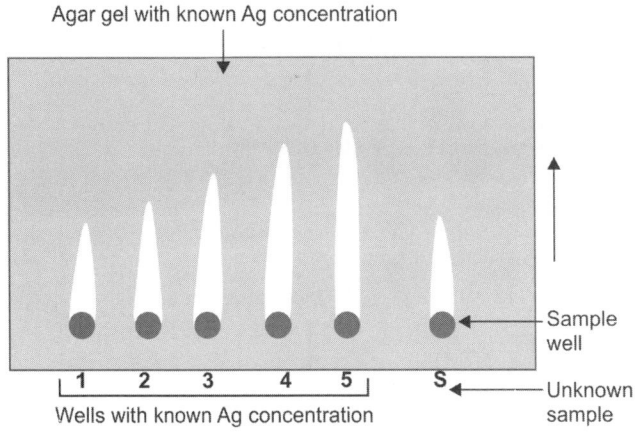

Fig. 4.7.5: Rocket immunoelectrophoresis.

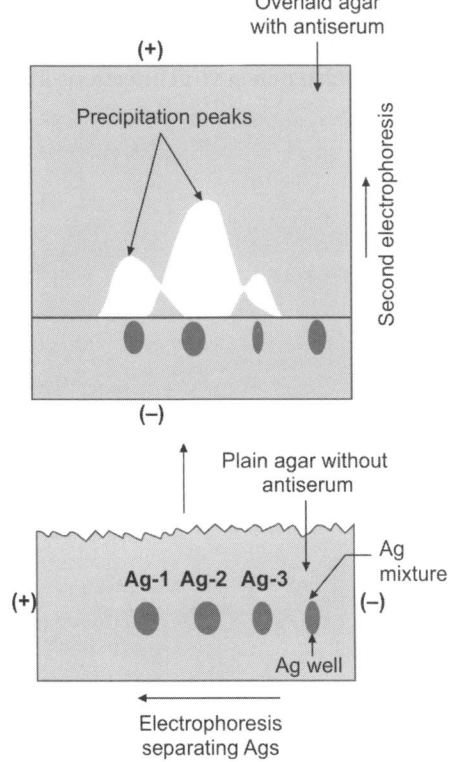

Fig. 4.7.6: Two-dimensional immunoelectrophoresis.

sufficiently. Poorly charged proteins like Igs are not properly analyzed and quantitated by this method (**Fig. 4.7.6**).

Agglutination Reaction:

❖ When the antigen is of particulate nature, its reaction with specific antibodies leads to formation of visible clumps which is called as agglutination. The antibodies producing agglutination are called agglutinins.

❖ These reactions are similar to precipitation reaction and require bivalent or polyvalent antigen and optimal Ab-Ag concentration.

1. Hemagglutination:

This is routinely performed in blood group matching. The RBCs of the donor and recipients are mixed with antibodies and blood group antigens A and B on slide. If the antigens are present on RBC surface, they clump together and agglutinate. Depending on the type of antigen present or absent, blood grouping is decided and matched.

2. Bacterial Agglutination:

❖ Used to detect bacterial infection. Bacterial infection leads to antibody formation.
❖ So, serum samples from patients are placed in increasing dilutions in different test tubes.
❖ The bacteria which is suspected to be the causative organism is then added to each test tube. The tube with maximum dilutions showing agglutination indicates the titer of antibody.

The agglutination titer is defined as reciprocal of **greatest serum dilution producing a positive agglutination**, e.g. in a serial dilution if 1:200 shows agglutination but 1:400 does not, then the agglutination titer is 200 [Excess of antibody inhibits agglutination—a phenomenon called **prozone effect** which may be due to: (i) Univalent binding with the antigen without cross-linking of antigen and (ii) Due to presence of incomplete Abs, i.e. those antibodies which bind to antigen but do not produce agglutination. These are usually of IgG isotypes and by binding to antigen, it prevent the binding of IgM which is a good agglutinin].

Typhoid is routinely diagnosed by agglutination reaction.

Passive/Indirect Agglutination:

❖ In this method, soluble antigen which normally do not cause agglutination are coated on RBCs by treating them with tannic acid or chromium chloride, both of which promote antigen adsorption on RBCs.
❖ Alternatively, synthetic beads are coated with antigen. These are then used to detect antibodies by agglutination reaction or vice versa as in bacterial or hemagglutination.

Agglutination inhibition assay:

❖ In this assays, absence of agglutination is the indicator of a positive result.
❖ The test samples containing the antigen is first incubated with specific antibody and then mixed with antigen-coated beads. There will be no clumping if the test sample contained antigen because it is bound

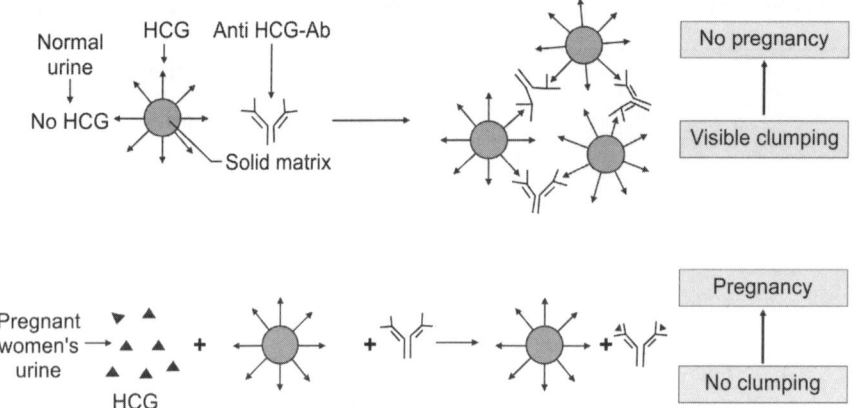

Fig. 4.7.7: Principle of pregnancy test based on agglutination inhibition.

by antibody. There is no free antibody left to bind to antigen-coated beads and cause their cross-linking.
- However, if sample did not contain antigen, then Ab will be available to cause clumping of beads.
- This was the basis of earlier available pregnancy test kits, which were based on detection of HCG in urine as marker of pregnancy **(Fig. 4.7.7)**.
- Similar assays also detect viral infections like rubella.

NOTES

4.8 Quality Control in Clinical Laboratory

> **LEARNING OBJECTIVES**
> At the end of session student must be able to:
> ⊃ Know importance, requirement, advantage and disadvantage of automation in clinical laboratory
> ⊃ Know about quality assurance programme runs in clinical laboratory.

Competency Bl.11.16: Observe use of commonly used techniques in biochemistry laboratory.
Domain: Knows.
Level: Knows how.
Core competency: Yes.

INTRODUCTION

Quality assurance in clinical biochemistry laboratory is a term usually applied to analytical performance and it incorporates the activities of external quality assessment and internal quality control.

A quality control system is designed to ensure that individual patient sample measurements are within clinically acceptable limits.

Process or system for monitoring the quality of laboratory testing, which ensure sensitivity and specificity of laboratory results and also determine the accuracy and precision of results.

PRECISION AND ACCURACY

Precision

It refers to the reproducibility of the test, i.e. the extent of variation of results obtained by repeated determination on a sample of the same estimation. If the variation is minimal, the method is precise. Precision reflects the correctness of carrying out the investigations.

Thus, if a particular test gives accurate and precise result, it means that the results obtained is quite close to the actual value and repeated determination will not change the value significantly.

In addition, two more concepts—specificity and sensitivity also determine the reliability of a result. Without going into the formal definition and details, these can be understood as below.

Accuracy

It is the net outcome of many quality parameters, i.e. accuracy, precision, specificity, sensitivity, etc. It can be assumed as a total reflection of the overall quality.

The results of laboratory have to be reliable to fulfill their role in diagnosis, treatment, prevention and screening, and understanding of diseases. Strict quality control program has to be maintained to ensure reliable results.

It refers to the closeness of the estimated value of an analyte to its actual value in a given sample. If repeated determinations are carried out on a sample of the same analyte, the closeness of the mean value of all such results to the actual value determines the assurance. This is dependent on methodology used.

SENSITIVITY AND SPECIFICITY

Sensitivity

It reflects the ability of a test to estimate even the minute amount of the target substance present in the sample. It is a reflection of the ability of detection of the positive result out of the total positive results.

So, if a test is specific, sensitive, accurate, and precise, then its results are within the reliable quality.

Specificity

It represents the ability of the test to discriminate between the similar substances that it intends to measure. A good example is estimation of blood glucose by reduction methods and enzymatic methods because glucose oxidase enzyme is highly specific, it only acts on glucose and not on other hexoses. Reduction methods are nonspecific because they use the reduction property of all sugars. Hence, if nonglucose contribution in the final result is not taken into account, some normal individuals may be wrongly labeled as diabetic (false positive). It is a reflection of the ability of detection of the negative results out of the total negative results by a test.

How to implement a laboratory quality control program?

- Select high quality controls
- Analyze the controls
- Perform statistical analysis
- Develop Levey-Jennings chart
- Monitor control values using the Levey-Jennings chart and/or Westgard rules
- Take immediate corrective action, if needed *Record actions are taken.*

CHARACTERISTICS OF CONTROLS

Values cover medical decision points and similar to the test specimen (matrix).

Available in large quantity and stored in small aliquots ideally, should last for at least 1 year. Often use biological material, consider biohazardous.

May be frozen, freeze-dried, or chemically preserved, requires very accurate reconstitution, if this step is necessary and always store as recommended by manufacturer.

TYPES OF CONTROLS

- *Assayed:* Mean calculated by the manufacturer
- *Unassayed:* Less expensive, must perform data analysis
- *"Homemade" or "In-house":* Pooled sera collected in the laboratory and preserved in small quantities for daily use.

ANALYSIS AND MONITORING OF QC

Run controls sample over a period of month: Run controls on daily basis with batches or between the batches to see that they continue to indicate the same performance. *For each parameter, there is requirement of 20 consecutive value of each level of control for measurement of variation of results.* A certain amount of variability will naturally occur when a control is tested repeatedly. Variability is affected by operator technique, environmental conditions, and the performance characteristics of the assay method. The goal is to differentiate between variability due to chance from that due to error.

Calculate mean, standard deviation, and develop Levey-Jennings charts, plot results, and calculate coefficient of variation (for each parameter).

Monitor over time to evaluate the precision and accuracy of repeated measurements. Review charts and documents at defined intervals, take necessary action (based on Westgard rules).

MONITORING OF QC PROGRAM

- *Internal quality control program:* Day-to-day, internal quality control program is very useful to evaluate the precision and accuracy of your results. Control sample can be prepared by the laboratory itself or use a commercial product, often which are based on animal sera to avoid danger of hepatitis infection.
- *External quality control program:* Once in a month, external quality control program is designed and aimed to provide comparison of your laboratory results with all participating laboratories.

NOTES

Quality Control in Clinical Laboratory

Exercise 4.8.1: Quality control in clinical laboratory and ATCOM sensitization.

1. Visit your hospital's central diagnostic laboratory and gather information and narrate your experience about internal/external quality assurance program of clinical biochemistry laboratory.

REFLECTIVE WRITING

NOTES

4.9 DNA Isolation from Blood and Tissue

> **LEARNING OBJECTIVE**
> At the end of session student must be able to know about principle, procedure and precautions for DNA isolation and quantification.

Competency BI.11.16: Observe use of commonly used techniques in biochemistry laboratory.
Domain: Knows.
Level: Knows how.
Core competency: Yes.
DNA—Isolation and quantification.

INTRODUCTION

DNA is the basis of life in living forms. DNA is the basic hereditary unit of each and every cell. Molecular biology is one of rapidly developing science related with understanding of processes of life at the DNA level. Molecular research involves study of abnormalities of structure and function of DNA (at the level of gene), which are the basis of various diseases. Extraction of DNA and its estimation is the first step in any investigation on DNA. Hence, it is important to be familiar with this basic step.

PRINCIPLE

DNA concentration is measured using a double cell UV spectrophotometer. The nucleic acid absorbs light strongly in the UV region at 260 nm due to conjugated bonds present in purine and pyrimidine bases.

STEPS

The cells are lysed using a detergent and then digested with proteinase K. The proteins are then separated from the cell by extracting repeatedly with phenol and chloroform.

Proteins being soluble in phenol/chloroform separate out as interphase between phenol and aqueous layer after centrifugation while DNA and RNA are contained in aqueous layer. RNA is removed by RNase treatment and DNA is precipitated out using ethanol and sodium acetate.

REAGENTS

- Tris-EDTA (TE) buffer—pH-8.0, containing 10 mM Tris, 1 mM EDTA
- DNA lysis buffer—3% SDS in 2x TE (pH-8.1)
- Proteinase K—stock 20 mg/mL in 20 mM Tris-HCl buffer; pH-7.5
- Sodium acetate—3 M
- Saturated phenol in TE—also contains 2 mg of 8-hydroxyquinoline in 100 mL of TE buffer
- Chloroform—isoamyl alcohol (24:1) (CIA).

PROCEDURE

1. Take a small pinch of tissue from biopsy sample of scrapping and mince in a mortar and pestle using 0.5 mL of 1x TE buffer and a pinch of sand.
2. Add 2 mL of 1x TE buffer and transfer the tissue homogenate to 15 mL falcon tube. Rinse mortar with further 1.5 mL of TE buffer.
3. Add lysis buffer in ratio of 1:2, i.e. approximately 50% of total volume of the homogenate.
4. Add such a volume of proteinase K (10 mg/mL) that in the final volume of the solution proteinase K is 100 μg/mL.

5. Incubate 37°C overnight.
6. Then add equal volume (~6 mL) of phenol and mix thoroughly in shaker for 15 minutes.
7. Centrifuge at 5,000 rpm for 10 minutes at 4°C.
8. Take supernatant in a fresh tube using Pasteur pipette.
9. Add an equal volume of phenol and CIA (1:1).
10. Shake in shaker for 15–30 minutes. Centrifuge and take supernatant in a fresh tube. Repeat step 9.
11. To the supernatant, add equal volume of CIA.
12. Shake for 15 minutes and then centrifuge at 5,000 rpm for 10 minutes at 4°C.
13. To the aqueous supernatant, add 1/10th volume of chilled 3 M sodium acetate solution and 2.5 volume of chilled absolute ethanol.
14. DNA forms a precipitate and can be seen as small coil that can be spooled out in an Eppendorf tube or pelleted by centrifugation.
15. Wash the DNA pellet with 70% ethanol and air dry.
16. Add 100–200 µL of 1x TE buffer to dissolve the pellet.
17. Store DNA at –20 to –70°C till use.

QUANTIFICATION OF DNA

- Make an accurate dilution of the DNA in distilled water and mix well.
- Fill three quarters of the reference cell of spectrophotometer with distilled water plus some TE buffer. Avoid cross-contamination and bubble formation in cells. Similarly, fill the sample cell with diluted DNA.
- Adjust spectrophotometer sequentially to measure OD at 260 nm and 280 nm.
- Note the OD value and calculate the purity and concentration of DNA.

PRECAUTIONS

- Stress should be avoided to prevent shearing of DNA. So, no vortex mixing or vigorous pipetting should be done.
- All glassware, plasticware, and reagents must be autoclaved before use.
- Phenol is corrosive, so it should be handled carefully.

NOTES

UNIT 5

Early Clinical Exposure Exercises and Reflective Writing

OUTLINES

5.1. Analysis of Cerebrospinal Fluid
5.2. Thyroid Function Test
5.3. Pancreatic Function Tests
5.4. Disorders of Acid-Base Balance

Analysis of Cerebrospinal Fluid

5.1

LEARNING OBJECTIVES

At the end of session student must be able to:
- Know and identify normal constituents of CSF
- Interpret and correlate the finding of CSF with various pathological conditions

Competency BI.11.15: Describe and discuss the composition of CSF.
Domain: Knows.
Level: Knows how.
Core competency: Yes.

CEREBROSPINAL FLUID

Brain and spinal cord are covered by three membranes—meninges. The dense fibrous outer dura mater, thin innermost pia mater, and trabeculated middle arachnoid mater. The last two are grouped as leptomeninges. The space between pia mater and arachnoid mater is subarachnoid space, which is filled with Cerebrospinal Fluid (CSF). The secretory activity of cells of the choroid plexus is the major factor in the production of CSF and ultrafiltration of plasma plays only a secondary role in CSF formation.

1. **Location:** Cerebrospinal fluid is present in subarachnoid space.
2. **Formation:** It is formed from choroid plexus of fourth ventricle of the brain.
3. **Collection:** Needle is inserted into subarachnoid space between 3rd and 4th lumbar vertebrae about 4 cm deep under all aseptic conditions.
4. **Functions:**
   - It serves as a fluid buffer against injury.
   - It acts as a reservoir to regulate the contents of cranium.
   - It serves as medium for nutrients exchange in the CNS.

Composition of Normal CSF

1. **Physical characteristics:**
   - *Amount:* The normal amount is 100–150 mL.
   - *Pressure:* Normal pressure of CSF in horizontal position varies between 100 mm H_2O and 200 mm H_2O.
   - *Appearance:* Clear, colorless, no coagulum, and no sediments.
   - *Reaction:* Alkaline (pH 7.3–7.5).
   - *Specific gravity:* 1.006–1.007
2. **Chemical composition:**
   - *Protein (mostly albumin):* 15–45 mg/dL
   - *Glucose:* 45–75 mg/dL
   - *Chlorides:* 120–130 mEq/L
   - *Calcium:* 5.5–6 mg/dL
   - *Inorganic phosphate:* 1.5–2.1 mg/dL
   - *Urea:* 20–40 mg/dL
   - *IgG:* 30 mg/mL
   - *IgA:* 4 mg/dL.
3. **Cells (mainly lymphocytes):** 0–4 cells/mm^3.

Biochemical examination of CSF is important in many diseases of brain, spinal cord, and meninges and is done routinely. Subarachnoid hemorrhage is diagnosed by the

Table 5.1.1: Characteristics of CSF in physiological and pathological conditions.

Clinical condition	Color and appearance	Cell count	Protein	Sugar	Coagulation
Normal	Clear and colorless	0–4×10^6/L	15–45 mg/dL	45–75 mg/dL	Not seen
Bacterial meningitis (purulent meningitis)	Opalescent or turbid due to high cell content	Markedly increased polymorphs	Marked increase	Marked decrease	May clot on standing
Tuberculous meningitis	May be opalescent	Lymphocytes and mononuclear cells	Increased	Low but not very much decreased	Cobweb-type coagulation
Viral infection	Clear and colorless	–	Increased	Normal	Nil
Brain tumor	Clear and colorless	Within normal range	Increased	Low	Solidifies
Subarachnoid hemorrhage	Blood stained in fresh hemorrhage	RBCs and WBCs	Increased	Not significant	Nil

presence of blood in CSF. Rise in protein content is seen in spinal tumor and in meningitis. Glucose level is increased in diabetes mellitus and decreased in meningitis. The chloride content is reduced in any condition associated with low blood chloride level such as vomiting, diarrhea, etc.

Indications of CSF Analysis

- *Infections:* Meningitis and encephalitis (bacterial, viral, or tubercular).
- *Trauma:* To diagnose subarachnoid hemorrhage.
- Tumors of brain
- Degenerative diseases of brain like multiple sclerosis.
- In cerebrovascular accidents.

Contraindications

- Raised intracranial tension
- Local skin infections at the site of puncture.

Fluid is collected in three different vacuettes:

1. *For biochemical tests:* Use fluoride (gray cap) and plain vacuette (red cap)
2. *For bacterial culture:* Use plain sterile vacuette (green cap)
3. *For cell count:* Use EDTA vacuette (lavender cap).

Clinical Significance

CSF is tested in neurological disorders such as:
- Subarachnoid hemorrhage
- Brain/spinal tumor
- Meningitis, multiple sclerosis
- Tuberculosis.

Meningitis

It is inflammation of meninges. It can be cerebral, spinal, or cerebrospinal. Pachymeningitis involves dura mater while leptomeningitis involves pia and arachnoid mater.

Encephalitis

Inflammation of brain parenchymal cells manifesting with changes in level of consciousness, increased intracranial pressure, and sensory motor dysfunction.

Poliomyelitis

Acute viral disease that causes destruction of anterior horn cells in spinal cord and often cranial nerve nuclei with ensuing palsy.

VIVA VOCE QUESTIONS: CEREBROSPINAL FLUID

1. Where CSF is present and what is the indication and procedure of CSF collection?
2. Mention characteristic of CSF and its clinical applications.

NOTES

Unit 5: Early Clinical Exposure Exercises and Reflective Writing

Exercise 5.1.1: Early clinical exposure and ATCOM sensitization.

Visit your hospital's pediatric ward and gather information about children those undergone CSF examination and try to find and correlate their laboratory finding and narrate your experience.

REFLECTIVE WRITING

NOTES

Thyroid Function Test

5.2

LEARNING OBJECTIVES
At the end of session student must be able to:
- Know about different laboratory tests and methods of assessment of thyroid functions
- Interpret the laboratory results related with dysfunction and disorders of thyroid
- Explain biochemical basis and need of lab test for thyroid disorders

Competency Bl.6.13: Describe the functions of the thyroid glands.
Competency Bl.6.14: Describe the tests that are commonly done in clinical practice to assess the functions of thyroid.
Domain: Knows.
Level: Knows how.
Core competency: Yes.

INTRODUCTION

Thyroid Gland

- Single, bilobed, and butterfly-shaped gland.
- Located in the anterior and lower midline part of the neck.
- Largest of all endocrine glands.
- Weighs 20–25 g.
- Unique among human endocrine glands—it stores large amounts of inactive hormone within extracellular follicles.
- Responsible for synthesis and secretion of T3 and T4. Occasional scattered "clear cells"/parafollicular cells/"C cells" produce and secrete **calcitonin**.

Regulation of Thyroid

The production of thyroxine and triiodothyronine is primarily regulated by thyroid-stimulating hormone (TSH), released by the anterior pituitary gland. TSH release, in turn, is stimulated by thyrotropin-releasing hormone (TRH), released in a pulsatile manner from the hypothalamus. The thyroid hormones provide negative feedback to TSH and TRH.

When the thyroid hormones are high, TSH production is suppressed.

When levels of TSH are high, TRH production is to be suppressed **(Fig. 5.2.1)**.

Thyroid Disorders

Hypothyroidism

- **Primary hypothyroidism:**
  A. *Defect in thyroid hormone synthesis:* Iodine deficiency, congenital, antithyroid agents, and drug—lithium.
  B. *Loss of functional tissue:* Hashimoto's thyroiditis, radiation, postoperative, and developmental.
  C. *Infiltrative disease:* Viral or bacterial infection of thyroid.
- **Secondary hypothyroidism:**
  Thyroid's inability to produce hormones due to malfunctioning of pituitary gland or hypothalamus.

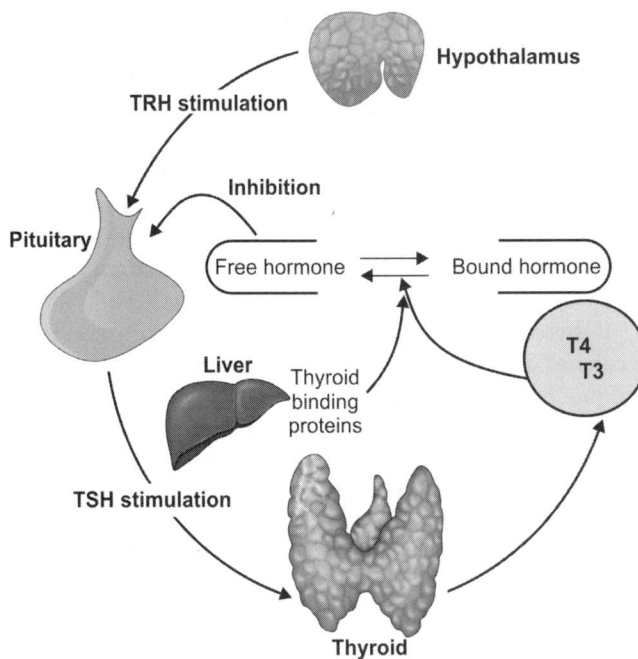

Fig. 5.2.1: Regulation of thyroid.

Hyperthyroidism

- **Primary hyperthyroidism:**
  - Graves' disease
  - Multinodular goiter
  - Autonomous nodule
  - Exogenous thyroid hormone
  - Transient—subacute thyroiditis, postpartum thyroiditis
  - Thyroid carcinoma (papillary, follicular, and anaplastic)
  - Drug—amiodarone.
- **Secondary hyperthyroidism:**
  Excess production of thyroid hormone due to malfunctioning of pituitary gland or hypothalamus.

	Normal	Primary hyperthyroidism	Primary hypothyroidism	Secondary hyperthyroidism	Secondary hypothyroidism
TSH	Normal	Low	High	High	Low
T4	Normal	High	Low	High	Low

Thyroid Function Test Panel

- TSH test
- T4 test and free T4 test
- T3 test and free T3 test
- Thyroid antibody test
- Thyroid ultrasound
- Thyroid scan
- Radioactive iodine uptake test.

TSH Test

The best way to initially test thyroid function is to measure the TSH level in a blood sample.

A high TSH level most often means hypothyroidism, or an underactive thyroid. This means that thyroid gland does not making enough hormone. As a result, the pituitary keeps making and releasing TSH into blood.

A low TSH level usually means hyperthyroidism, or an overactive thyroid. This means that thyroid gland is making too much hormone, so the pituitary stops making and releasing TSH into your blood.

If the TSH test results are not normal, it will need at least one other test to help find the cause of the problem.

T4 Test and Free T4 Test

A high blood level of T4 may mean: hyperthyroidism.
A low level of T4 may mean: hypothyroidism.

T4 circulates in the blood in two forms: (1) T4 bound to proteins that prevent the T4 from entering the various tissues that need thyroid hormone and (2) Free T4, which does enter the various target tissues to exert its effects. The free T4 fraction is the most important to determine how the thyroid is functioning, and tests to measure this are called the *free T4 (FT4)* and the *free T4 index (FT4I or FTI)*.

Individuals who have hyperthyroidism will have an elevated FT4 or FTI, whereas patients with hypothyroidism will have a low level of FT4 or FTI. Combining the TSH test with the FT4 or FTI accurately determines how the thyroid gland is functioning. The finding of an elevated TSH and low FT4 or FTI indicates primary hypothyroidism due to disease in the thyroid gland. A low TSH and low FT4 or FTI indicates hypothyroidism due to a problem involving the pituitary gland. A low TSH with an elevated FT4 or FTI is found in individuals who have hyperthyroidism.

T3 Test and Free T3 Test

T3 tests are often useful to diagnose hyperthyroidism or to determine the severity of the hyperthyroidism. Patients who are hyperthyroid will have an elevated T3 level. In some individuals with a low TSH, only the T3 is elevated and the FT4 or FTI is normal.

Thyroid Antibody Test

The immune system of the body normally protects us from foreign invaders such as bacteria and viruses by destroying these invaders with substances called antibodies produced by blood cells known as lymphocytes. In many patients with hypothyroidism or hyperthyroidism, lymphocytes make antibodies against their thyroid that either stimulate or damage the gland. Two common antibodies that cause thyroid problems are directed against thyroid cell proteins: **thyroid peroxidase and thyroglobulin**.

Measuring levels of thyroid antibodies may help diagnose the cause of the thyroid problems.

Positive antithyroid peroxidase and/or antithyroglobulin antibodies in a patient with hypothyroidism make a diagnosis of *Hashimoto's thyroiditis*.

If the antibodies are positive in a hyperthyroid patient, the most likely diagnosis is *autoimmune thyroid disease*.

Thyroid Ultrasound

Ultrasound of the thyroid is most often used to look for thyroid nodules. Thyroid nodules are lumps in neck. Ultrasound can helpful to differentiate whether thyroid nodule is benign or cancerous.

Thyroid Scan

Thyroid scan is useful to look at the size, shape, and position of the thyroid gland. This test uses a small amount of radioactive iodine to help find the cause of hyperthyroidism and check for thyroid nodules.

Radioactive Iodine Uptake Test

A radioactive iodine uptake test, also called a thyroid uptake test, is helpful to find the cause of hyperthyroidism. The thyroid "takes up" iodine from the blood to make thyroid hormones, which is why this is called an uptake test.

If thyroid collects a large amount of radioactive iodine, it can be Graves' disease, or one or more nodules that make too much thyroid hormone.

Even though the test uses a small amount of radiation and is thought to be safe, yet these tests are contraindicated in pregnancy and during breastfeeding.

VIVA VOCE QUESTIONS: THYROID FUNCTION TEST

1. How T4 and T3 are synthesized?
2. What is the importance of total T4 and T3 as compared to FT4 and FT3? Which is better for evaluation of thyroid?
3. What are the common causes of hypothyroidism?
4. Name the common factors, which alter the free T4 and T3 levels by interference in their binding with proteins.

NOTES

Unit 5: Early Clinical Exposure Exercises and Reflective Writing

Exercise 5.2.1: Early clinical exposure and ATCOM sensitization.

Visit to hospital central diagnostic laboratory and try to find patients with thyroid disorders with the help of your teachers, interview them to find signs and symptoms, and try to correlate clinical finding with laboratory results.
Narrate your experience.

REFLECTIVE WRITING

NOTES

Pancreatic Function Tests

5.3

> **LEARNING OBJECTIVES**
> At the end of session student must be able to:
> ⊃ Know about different laboratory tests and methods of assessment of pancreatic functions
> ⊃ Interpret the laboratory results related with dysfunctions and disorders of pancreas
> ⊃ Explain biochemical basis and need of lab test for pancreatic disorders.

Competency PY.4.8 and PY.8.4: Describe and discuss pancreatic function tests.
Domain: Knows.
Level: Knows how.
Core competency: Yes.

PANCREATIC PROFILE

Pancreas has two functions:
1. As an endocrine gland that synthesizes hormones like insulin and glucagon.
2. As an exocrine gland by synthesizing potent digestive enzymes like trypsin, chymotrypsin, elastase, carboxypeptidase, lipase, etc. to facilitate duodenal digestion.

PANCREATIC DISEASES (EXOCRINE)

A. **Examination of duodenal contents following pancreatic stimulation:** Estimate the output of the fluids, bicarbonate, and enzymes, which can be stimulated by hormones like secretin, cholecystokinin, or pancreozymin indirectly by a test meal.
   - **Stimulation using exogenous hormones:**
     - **Secretin stimulation test:**
       Positioning of tube for sample collection.
       Insertion of double lumen tube with one part 2.5 cm longer than other and is positioned in specific region of GI tract by avoiding any contamination to biliary, salivary, and gastric secretion.
       Collection of basal endogenous duodenal contents is done after 10 minutes of insertion of tube.
       Stimulation of pancreas is done by giving intravenous secretin (1 U/kg).
       Collection of samples after 10-minute interval has been done for measurement of volume, bilirubin, bicarbonate, and enzymes.
       This test is helpful in diagnosis of chronic pancreatitis, pancreatic carcinoma, etc.
     - **Cholecystokinin-stimulation Test:**
       Pancreozymin is given immediately after the collection of sample for the secretin stimulation test.
       This test is helpful to assess pancreatic function.
   - **Indirect stimulation of pancreas (Lundh test):**
     Duodenal contents, specifically trypsin, are analyzed after an oral standard test meal.

B. **Indirect study methods for pancreas:**
   - **Tests based on determination of enzymes in serum and urine: estimation of serum amylase and serum lipase**—levels are elevated in acute pancreatitis due to damaged cells of pancreas pass into blood. While in obstructive diseases like tumors, calculus, and fibrous tissue formation, rise in enzymes levels are to a lesser extent.
   - **Test based on impaired digestion:** Average daily fecal excretion of fat measurement is done to detect progress of established pancreatic disorders.

- ➢ **Test based on impaired carbohydrate metabolism:** GTT is done to detect involvement of endocrine tissues of the pancreas.
- ➢ **Other laboratory tests in acute pancreatitis:**
  - ♦ Plasma glucose—to detect transient hyperglycemia
  - ♦ Serum bilirubin—increase in alcohol-related pancreatitis
  - ♦ Serum trypsin, serum lecithinase, and deoxyribonuclease—elevated in pancreatitis
  - ♦ Serum calcium—fall in calcium level in pancreatic disorders
  - ♦ WBC count—leukocytosis
  - ♦ Hb—hemoconcentration (raised Hb)
  - ♦ Serum lipid profile—hyperlipidemia.

  In fibrocystic diseases affecting pancreas, high concentration of Na⁺ and Cl⁻ ions is seen in sweat.

- ➢ **Visual procedures:** Ultrasonic scanning, computed tomographic scanning, and endoscopic retrograde cholangiopancreatography.

VIVA VOCE QUESTIONS: PANCREATIC FUNCTION TESTS

1. Write clinical implication of pancreatic function tests.
2. What is the role of endocrine pancreas in human body?
3. What is the role of insulin in gluconeogenesis?
4. Mention the laboratory investigations and its clinical correlations in acute pancreatitis.
5. What is the importance of serum amylase and serum lipase estimation in patients with acute pancreatitis?

NOTES

Exercise 5.3.1: Early clinical exposure and ATCOM sensitization.

Visit to hospital gastroenterology/medicine department and try to find patients with pancreatic disorders with the help of your teachers, interview them to find signs and symptoms, and try to correlate clinical finding with laboratory results.

Narrate your experience.

REFLECTIVE WRITING

NOTES

5.4 Disorders of Acid-Base Balance

> **LEARNING OBJECTIVES**
> At the end of session student must be able to:
> ⊃ Know about homeostasis mechanism for acid-base balance regulation in body
> ⊃ Know about different disorders related with acid-base imbalance in body

Competency BI.11.17: Explain the basis and rationale of biochemical tests done in the acid-base imbalance.
Domain: Knows.
Level: Knows how.
Core competency: Yes.

INTRODUCTION

Acid-base homeostasis involves chemical and physiologic processes responsible for the maintenance of the acidity of body fluids at levels that allow optimum operation of many cellular enzymes and function of vital organs, predominantly the brain and the heart.

Acids are substances that are capable of donating protons.

Physiologically important acids include:
- **Carbonic acid (H_2CO_3)**
- **Phosphoric acid (H_3PO_4)**
- **Pyruvic acid ($C_3H_4O_3$)**
- **Lactic acid ($C_3H_6O_3$)**

These acids are dissolved in body fluids.
Bases are those that accept protons.

Physiologically important bases include:
- **Bicarbonate (HCO_3^-)**
- **Biphosphate (HPO_4^{2-})**.

ACID-BASE REGULATION

The pH of plasma is 7.4. In normal life, the variation of plasma pH is very small. The pH of plasma is maintained within a narrow range of 7.35–7.45.

The body has developed three lines of defense to regulate the body's acid-base balance and maintain the blood pH (around 7.4).

THE THREE–TIER DEFENCE

First line of defence: Blood buffers
Second line of defence: Respiratory regulation
Third line of defence: Renal regulation.

Blood Buffers

Buffers are solution, which resists any change of pH. It is a mixture of weak acid and its salt with a strong base/mixture of weak base and its salt with strong acid. Buffering capacity of a buffer is defined as the ability of the buffer to resist change in pH when an acid or base is added.

Types of blood buffers:
- Sodium bicarbonate buffer
- Phosphate buffer
- Protein buffer/hemoglobin buffer.

Buffers act quickly, but not permanently. Buffers can respond immediately to addition of acid or base, but they do not serve to eliminate the acid from the body.

Respiratory Regulation of Acid-base Balance

The respiratory system contributes to the balance of acids and bases in the body by regulating the blood levels of carbonic acid. CO_2 in the blood readily reacts with water to form carbonic acid, and the levels of CO_2 and carbonic acid in the blood are in equilibrium. When the CO_2 level in the blood rises (as it does when you hold your breath), the excess CO_2 reacts with water to form additional carbonic acid, lowering blood pH. Increasing the rate and/or depth of respiration (which you might feel the "urge" to do after holding your breath) allows you to exhale more CO_2. The loss of CO_2 from the body reduces blood levels of carbonic acid and thereby adjusts the pH upward, toward normal levels. As you might have surmised, this process also works in the opposite direction. Excessive deep and rapid breathing (as in hyperventilation) rids the blood of CO_2 and reduces the level of carbonic acid, making the blood too alkaline. This brief alkalosis can be remedied by rebreathing air that has been exhaled into a paper bag. Rebreathing exhaled air will rapidly bring blood pH down toward normal.

The body regulates the respiratory rate by the use of chemoreceptors, which primarily use CO_2 as a signal. Peripheral blood sensors are found in the walls of the aorta and carotid arteries. These sensors signal the brain to provide immediate adjustments to the respiratory rate if CO_2 levels rise or fall, yet other sensors are found in the brain itself. Changes in the pH of CSF affect the respiratory center in the medulla oblongata, which can directly modulate breathing rate to bring the pH back into the normal range.

Abnormally elevated blood levels of CO_2 (hypercapnia) occurs in any situation that impairs respiratory functions, including pneumonia and congestive heart failure. Reduced breathing (hypoventilation) due to drugs such as morphine, barbiturates, or ethanol (or even just holding one's breath) can also result in hypercapnia.

Abnormally low blood levels of CO_2 (hypocapnia) occurs with any cause of hyperventilation that drives off the CO_2, such as salicylate toxicity, elevated room temperatures, fever, or hysteria.

Renal Regulation

Four mechanisms:
1. Excretion of H^+
2. Reabsorption of bicarbonate
3. Excretion of titratable acids
4. Excretion of ammonium ion.

ACID-BASE DISORDERS

Acidosis and alkalosis are not diseases, but rather are the results of a wide variety of disorders.

Acidosis

If pH is below 7.38, it is known as acidosis. Life is threatened when pH is lowered below 7.25. Acidosis leads to CNS depression and coma.
- Respiratory acidosis
- Metabolic acidosis.

Respiratory acidosis: Primarily increase in pCO_2.

Causes of respiratory acidosis: Pneumonia, bronchitis, asthma, and chronic obstructive lung disease, pneumothorax, sedatives, peritonitis, and sleep apnea, paralysis of respiratory muscles, narcotics.

Metabolic acidosis: Primarily decrease in bicarbonate levels.

Causes of metabolic acidosis: Diarrhea, intestinal fistula, renal tubular acidosis, carbonic anhydrase inhibitor—acetazolamide, ureterosigmoidostomy, and drugs—antacids containing Mg, iodide (absorbed from dressings), Li, and polymyxin B.

Alkalosis

When pH is more than 7.42, it is alkalosis. It is very dangerous if pH is increased above 7.55. Alkalosis induces neuromuscular hyperexcitability and tetany.
- Respiratory alkalosis
- Metabolic alkalosis.

Respiratory alkalosis: Primarily decrease in pCO_2.

Causes of respiratory alkalosis:
High altitude, hyperventilation, and hysterical overbreathing, septicemia, meningitis, assisted breathing, febrile conditions, and congestive cardiac failure.

Metabolic alkalosis: Primarily increase in bicarbonate levels.

Causes of metabolic alkalosis:
- Severe vomiting, loop diuretics, hypovolemia, and pyloric stenosis
- Cushing syndrome, milk-alkali syndrome.

VIVA VOCE QUESTIONS: ACID-BASE BALANCE

1. What is normal pH of blood? Explain the role of blood buffers and respiratory mechanism in the maintenance of acid-base balance of the body.
2. What is anion gap? Enumerate the cause of normal anion gap acidosis.

NOTES

Unit 5: Early Clinical Exposure Exercises and Reflective Writing

Exercise 5.4.1: Early clinical exposure and ATCOM sensitization.

Visit to hospital emergency/ICU department and try to find patients with acid-base disorders with the help of your teachers, and try to correlate clinical findings with laboratory results.

Narrate your experience.

REFLECTIVE WRITING

NOTES

UNIT 6

Attitude, Ethics and Communication (AETCOM) Modules

OUTLINES

6.1. Introduction of Clinical Methods
6.2. What does it Mean to be a Doctor?
6.3. What does it Mean to be a Patient?
6.4. The Doctor-Patient Relationship
6.5. The Foundations of Communications

Introduction of Clinical Methods

6.1

LEARNING OBJECTIVES

At the end of session student must be able to:
- Know about effective communication skills
- Know and understand about effective and structured way of clinical interview with patients

It is a multi-step organized approach towards remedial of the patients in empathetic and re-assuring manner, which involves:
- Communication skills: The art of history taking
- Clinical examination
- Interpretation of clinical and laboratory data
- Management of disease/disorder.

Module-1: Effective Communication Skills: The C-L-A-S-S Protocol

There is a structured protocol for effective Clinical Interviews (five key steps)

C-Context	Physical set up of the area you choose for the interview
L-Listening skills	How to be an effective listener
A-Acknowledge	How to validate, explore and address emotions and concerns
S-Strategy	How to provide a management plan that the patient can understand
S-summary	How to summarize and clarify the conversation ensuring comprehension

1. C-Context (setting)

A private area with no distractions

Physical Space
- Choose an area where you can have a private conversation.
- Your eyes should be at the same level as the patient and/or family member (sit down if you need to).
- There should be no physical barriers between you.
- If you are behind a desk, have the patient and/or family members sit across the corner.
- Have a box of tissues available.

Family Members/Friends
- The patient should be seated closest to you.

Body Language
- Present a relaxed demeanour
- Maintain eye contact except when the patient becomes upset.

Touch
- Only touch a non-threatening area (hand or forearm).
- Be aware of cultural issues that may not allow touching.

2. L-listening skills

Be an effective listener.
Open Ended Questions
- "How did you manage with the new treatment?"
- "Can you tell me more about your concerns?"
- "How have you been feeling?"

Facilitating

- Allow the patient to speak without interruption.
- Nod to let patient know that you are following.
- Repeat a key word from the patient's last sentence in your first sentence.

Clarifying

- "So, if I understand you correctly, you are saying…"
- "Tell me more about that."

Time and Interruptions

- If there are time constraints, let the patient know ahead of time.
- Pagers and phone calls – don't answer, but if you must, apologize to the patient before answering.
- Try to prepare the patient if you know you will be interrupted.

3. A-Acknowledge Emotions

Explore, identify, and respond to the emotion.

The Empathic Response
- Identify the emotion.
- Identify the cause of the emotion.
- Respond by showing you have made the connection between the emotion and the cause.
  "That must have felt terrible when…" "Most people would be upset about this."
- You don't have to have the same feelings as the patient.
- You don't have to agree with the patient's feelings.

4. S-Strategy

Propose a plan that the patient will understand.

The Plan
- Appraise in your mind or clarify with the patient's expectations of treatment and outcome.
- Decide what the best medical plan would be for the patient.
- Recommend a strategy on how to proceed.
- Evaluate the patient's response.
- Collaborate and agree on the plan.

5. S-Summary

Closing the interview.

Final Thoughts
- Summarize the discussion in a clear and concise manner.
- Check the patient's understanding.
- Ask if the patient has any other questions for you.
- If you don't have time for further questions, suggest that he can be addressed at the next appointment.
- Make a clear contract for a follow up visit.

SUGGESTED READING

1. Interpersonal Communication and Relationship Enhancement (I*CARE) Program MD Anderson Cancer Centre – Faculty and Academic Development, University of Texas.

6.2 What does it Mean to be a Doctor?

LEARNING OBJECTIVES
At the end of this session each student should be able to know:
- Privileges and the responsibilities of the profession.
- What are the expectations of society from them?
- What will they have to do and give up in order meeting their own and society's expectations?

Competencies
1. Enumerate and describe professional qualities and roles of a physician.
2. Describe and discuss the commitment to lifelong learning as an important part of physician growth.
3. Describe and discuss the role of a physician in health care system.
4. Identify and discuss physician's role and responsibility to society and the community that she/he serve.

Level: Knows how.t

EXPLORATORY SESSION

Patients must be able to trust doctors with their lives and health. To justify trust, you must show respect for human life and make sure your practice meets the standards expected from you in following four domains:

1. Knowledge, Skill and Performance

- Make the care of your patient your first concern.
- Provide a good standard of practice and care.
- Keep your professional knowledge and skills up to date.
- Recognize and work within the limits of your competence.

2. Safety and Quality

- Take prompt action if you think that patient's safety, dignity or comfort is being compromised.
- Protect and promote the health of patients and the public.

3. Communication, Partnership and Teamwork

- Treat patients as individuals and respect their dignity.
- Treat patients politely and considerately.
- Respect patients' right to confidentiality.
- Work in partnership with patients.
- Listen to, and respond to, their concerns and preferences.
- Give patients the information they want or need in a way they can understand.
- Respect patients' right to reach decisions with you about their treatment and care.
- Support patients in caring for themselves to improve and maintain their health.
- Work with colleagues in the ways that best serve patients' interests.

4. Maintaining Trust

- Be honest and open and act with integrity.
- Never discriminate unfairly against patients or colleagues.

- Never abuse your patients' trust in you or the public's trust in the profession.
- You are personally accountable for your professional practice and must always be prepared to justify your decisions and actions

SUGGESTED READING

1. https://www.gmc-uk.org/ethical-guidance/ethical-guidance-for-doctors/good-medical-practice/duties-of-a-doctor.

What does it Mean to be a Patient?

6.3

> **LEARNING OBJECTIVES**
> At the end of this session, each student should be able to know:
> ⊃ The feelings and experiences of patients during their hospital stay
> ⊃ How does it affect patient's behavior, outlook and expectations?

AETCOM Competency:
1. Enumerate and describe professional qualities and roles of a physician.
   Level: Knows how.
2. Demonstrate empathy in patient encounters.
   Level: Shows how.

Module 1: When a Medical Student Becomes the Patient

First year student found lump in the base of his neck which results in a series of doctor visits, scans, discrepancies in the diagnosis, nail-biting waits for biopsies and finally received the conclusive biopsy result. The diagnosis: primary mediastinal B-cell lymphoma.

Action and Attitude: For Him—Hearing the word "cancer" outside the classroom, and in association with his body, was unbelievably difficult. He knows attitude can play an enormous role in healing. He would need chemotherapy and a fighting spirit. His life became his classroom and he was more motivated than ever to learn and to conquer his disease.

Finding Concern and Care: He realizes that a bit of compassion and patience is needed to cross the thin line between crippling fear and uplifting hope. Physicians hold that power in every interaction, balancing humanity and warmth with medicine and science.

The Path Ahead

He is treated with chemotherapy for 5 months. His post-chemo PET scans showed no trace of disease: complete remission—the most positive outcome. His unspeakable fear has been replaced by unshakable positivity. Beside that he achieved a major goal, having successfully finished his first year of medical school while learning more than any medical student should ever have to learn about disease.

QUESTIONS

1. Why it is important to understand patient's feeling and experience regarding his disease?
2. Write your reflection.

Unit 6: Attitude, Ethics and Communication (AETCOM) Modules

What happened?

....
....
....
....

So what

....
....
....
....

What next?

....
....
....
....

SUGGESTED READING

1. http://blogs.einstein.yu.edu/when-a-medical-student-becomes-the-patient.

The Doctor-Patient Relationship

6.4

LEARNING OBJECTIVES
At the end of this session, student should be able to:
- Describe the duties of a physician to his patient, public, colleagues and himself.
- Describe the significance of doctor-patient relationship.
- Describe the precincts of doctor-patient relationship.

Competency
1. Enumerate and describe professional qualities and roles of a physician.
   Level: Know how.
2. Demonstrate empathy in patient encounters.
   Level: Show how.

Module 1: Case from AETCOM Module/NMC

A 53-year-old man is seen by a cardiologist for chest pain lasting for a few minutes on accustomed exercise for the past 3 weeks. After a detailed history and physical examination, the doctor orders an ECG which was normal. He further orders an exercise stress test which showed reversible ischemia. The doctor orders an angiogram. At the time, the patient requests that he would like to have a second opinion. The cardiologist explains that he has done everything correctly and that the patient indeed requires an angiogram. The patient tells him that he cannot make a decision unless he talks to his family doctor of 20 years. The cardiologist is offended and tells the patient that he does not wish to see the patient any longer.

QUESTIONS
1. What is the importance of trust in the doctor-patient relationship?
2. What are the rights of a patient?
3. What are duties of a treating physician?
4. Does the request for a second opinion provide sufficient grounds to terminate the doctor-patient relationship?

Module 2: Case from AETCOM Module/NMC

A young doctor has been taking care of an 86-year-old woman for the past 2 years. She had a fall 2 years ago and has been mostly bedridden. She lives alone with just a care taker and her children are abroad. She requires preventive care mostly and the doctor makes house visits once a week. The doctor spends time talking to her during each visits and makes her feel comfortable. One day during such a visit, the patient expresses the view that her children have been ungrateful to her and that she intends to call her lawyer today and divide her assets between the doctor and the caretaker after her death. What should the doctor do?

QUESTIONS

1. What are the boundaries of a physician in the doctor-patient relationship?
2. Importance of trust and vulnerability in doctor-patient relationship.

Module 3: Panel Discussion/Symposium

A. Visit the following links
   https://www.nmc.org.in/rules-regulations/code-of-medical-ethics-regulations-2002.
B. A planned teaching learning session (role play) focusing on all components of doctor-patient relationship followed by reflective writing

What happened?

....
....
....
....
....
....
....

So what

....
....
....
....
....
....
....
....

What next?

....
....
....
....
....
....
....

SUGGESTED READING

1. https://www.nmc.org.in/wp-content/uploads/2020/01/AETCOM_book.pdf

6.5 The Foundations of Communications

> **LEARNING OBJECTIVES**
>
> At the end of this session, each student should be able to know:
> ⊃ The principle of communication.
> ⊃ The importance of effective communication for better doctor-patient relationship.
> ⊃ Teach basic communication and counselling skills to undergraduate students to increase their clinical competence.

Competency: Demonstrate ability to communicate to patients in a patient, respectful, nonthreatening, non- judgmental and empathetic manner.
Level: Shows how.

Module 1: Introduction of Communication Skills

Communication is both an art and a science. Effective communication is the cornerstone of patient-centred medicine and empathic behaviour, leading to a fruitful patient–physician relationship. It contributes to a positive therapeutic effect and better patient outcomes and satisfaction, thus increasing the overall quality of health care systems

Components to effective communication in a healthcare:
* *Conversational skill*: It involves focusing, clarifying, providing information, relevant intermittent questions, sharing humor, summarizing.
* *Listening skill*: It involves active listening with purpose, with silence, with acknowledgement of message.
* *Technical skill*: It involves sharing empathy, feeling, observations, non-verbal clues.
* To sum up effective communication should be clear, complete, concise and courteous.

Module 2: Doctor-patient Communication

A: Role play on doctor-patient communication (https://www.youtube.com/watch?v=S4wWClQhZaA)
B. A planned teaching learning session (role play) focusing on all components of effective communication followed by assessment based on Kalamazoo consensus

Kalamazoo consensus statement (Rating 1-3 Poor, 4-6 Satisfactory, 7-10 Superior)

Criteria	Score
Builds Relationship	
Opens Discussion	
Gathers Information	
Understands the Patients	
Shares Information	
Manages Flow	
Overall Rating	

Module 3: Reflective Writing

Visit of undergraduate students in small groups at hospital OPD setup to observe actual doctor-patient communication followed by Reflective writing.

What happened?

....
....
....
....
....
....
....

So what?

....
....
....
....
....
....
....
....

What next?

....
....
....
....
....
....
....
....

SUGGESTED READING

1. https://hsrc.himmelfarb.gwu.edu.Kalamazoo%20Essential%20Elements%20Communication%20Checklist.pdf

UNIT 7

Biochemical Calculations and Reference Range

OUTLINES

7.1. Preparations of Buffers and Solutions
7.2. Reference Value of Various Biochemical Parameters Integration with Medicine

7.1 Preparations of Buffers and Solutions

> **LEARNING OBJECTIVES**
> At the end of session student must be able to:
> ⊃ Know that how biochemical solutions and buffers are prepare
> ⊃ Prepare buffer solution for different biochemical techniques
> ⊃ Interpret and estimate pH of different solutions.

Competency BI.11.2: Describe the preparation of buffers and estimation of pH.
Domain: Knows.
Level: Knows how
Core competency: Yes.

Commonly used *units* **that a medical person should know:**
1. **Percent:** Ways of expressing percentage composition of a solution are:
   - **Weight per unit weight (W/W):** 10% (w/w) solution contains 10 g of solute in 90 g of solvent.
   - **Weight per unit volume (W/V):** This is the most commonly used unit. Its various forms are mg/dL, g/dL, and g/L.
     10% (w/v) solution contains 10 g of solute per 100 mL of solution.
   - **Volume per unit volume (V/V):** 10% (v/v) contains 10 mL of concentrate per 100 mL of solution.
2. **Molarity:** A molar solution (M) contains 1 g molecular weight (mole) of solute in 1 L of solution.
   - **Moles/unit volume** (e.g. mmol/L, μmol/L, moles/L, etc.):
     This is the most popular mode of expressing the concentration of a substance in biological fluids in medical literature all over the world.
     1 Mole of a substance refers to its molecular weight expressed in grams.

(1) The molecular weight of glucose is 180.
   1 mole of glucose = 180 g
   1 molar glucose solution is 180 g/L solution of glucose, prepared by dissolving 180 g of glucose in 1 L distilled water.
   1 mmol/L of glucose means a liter of solution containing 1 millimole.
   [1/1,000 (10^{-3}) of a mole or 180 mg] of glucose.
   - To convert mg/dL to mmol/L, the following formula is used:

$$\text{mmol/L of substance} = \frac{\text{mg/dL} \times 10}{\text{Molecular weight of substance}}$$

3. **Normality:**
   A normal solution contains 1 g equivalent weight of the solute in 1 L distilled water.
   - **Equivalents per volume** (mEq/L): It is the common unit for electrolyte concentration.
     mEq = mmol/valency

$$\text{mEq/L} = \frac{\text{mg/dL} \times 10}{\text{Equivalent weight}}$$

QUESTIONS

1. How would you prepare a liter of 0.5 M solution of glucose?
2. How much NaCl would you require to make 250 mL of molar solution of NaCl? (weight of Na = 23, chloride = 35.5)
3. How would you prepare 250 mL of a 100 mmol (100 millimolar) solution of creatinine? (molecular weight of creatinine is 113)
4. How would you prepare 250 mL of a 100 mmol (100 millimolar) solution of urea? (molecular weight of urea is 60)
5. Please express a 1 M/L of glucose as mg/dL.
6. Please express a 1 M/L of sucrose ($C_{12}H_{22}O_{11}$) as mg/dL. (atomic weight of C = 12, H = 1, and O = 16)
7. Please express 100 mg/dL solution of glucose as mmol/L.
8. Please express 100 mg/dL solution of creatinine as mmol/L.
9. How do you differentiate a blow-out pipette from a nonblow-out pipette?
10. Which finger is used to close the top end of a glass pipette?

7.2 Reference Value of Various Biochemical Parameters Integration with Medicine

Laboratory parameters	Reference range	Units
Diabetic Tests		
Blood sugar		
Fasting	70–100	mg/dL
Postprandial	70–140	mg/dL
Random	70–130	mg/dL
Oral GTT		
Fasting	>92	mg/dL
1 hour	>180	mg/dL
2 hours	>153	mg/dL
HbA1c		
Nondiabetic	<6%	
Excellent control	6–7%	
Good control	7–8%	
Fair control	8–9%	
Poor control	>9%	
Kidney Function Tests		
Urea	10–45	mg/dL
Creatinine	0.4–1.2	mg/dL
Electrolytes		
Sodium	136–145	mmol/L
Potassium	3.5–5.1	mmol/L
Chloride	97–111	mmol/L
Uric acid	3.5–7.2	mg/dL
BUN	6–21	mg/dL
Liver Function Tests		
Bilirubin		
Total	0.2–1	mg/dL
Direct	0–0.25	mg/dL
Indirect	0–0.75	mg/dL
SGOT	10–45	IU/L
SGPT	10–45	IU/L

Laboratory parameters	Reference range	Units
Alkaline phosphatase	Up to 85	IU/L
Protein		
Total protein	6.6–8.3	g/dL
Albumin	3.5–5	g/dL
Globulin	2.3–3.5	g/dL
A/G ratio	1.2–1.5	–
Lipid Profile		
Cholesterol	130–250	mg/dL
Triglycerides	0–170	mg/dL
HDL	30–70	mg/dL
LDL	0–150	mg/dL
VLDL	0–35	mg/dL
Cardiac Tests		
CPK total	46–171	IU/L
CPK-MB	0–25	IU/L
LDH	200–400	IU/L
TROP-T	Positive/Negative	–
Minerals		
Calcium	8–10	mg/dL
Phosphorus	2.5–5	mg/dL
Magnesium	1.5–2.6	mg/dL
Iron Profile		
Serum Iron		
Males	60–160	µg/dL
Females	35–145	µg/dL
Neonates	150–220	µg/dL
Ferritin	15–150	ng/mL
TIBC	250–400	µg/dL
Pancreatic Profile		
Amylase	Up to 85	IU/L
Lipase	Up to 85	IU/L

Laboratory parameters	Reference range	Units
Thyroid Profile		
T3	0.8–2	ng/mL
T4	4.6–11	mcg/dL
TSH	0.4–4.2	µIU/mL
Free T3	2.0–4.0	pg/mL
Free T4	0.93–1.7	ng/dL
Infertility Profile		
LH		
Males	3–15	mU/mL
Females		
Follicular	3–22	mU/mL
Midcycle	30–250	mU/mL
Postmenopausal	>30	mU/mL
FSH		
Males	5–15	mU/mL
Females		
Follicular	5–20	mU/mL
Midcycle	30–50	mU/mL
Postmenopausal	>35	mU/mL
Prolactin		
Males	4.04–15.2	ng/mL
Females	4.79–23.3	ng/mL
Anti-Müllerian hormone	1–6	ng/mL
Estradiol	21–251	pg/mL
Tumor Markers		
Carcinoembryonic antigen	0–5	ng/mL
Cancer antigen-125	0–135	IU/mL
Prostate-specific antigen	0–4	ng/mL

Laboratory parameters	Reference range	Units
Alpha-fetoprotein	<8.5	ng/mL
Vitamin B12		
Newborn	160–1,300	pg/mL
Adults	200–835	pg/mL
>60 years	110–800	pg/mL
Vitamin D		
Deficiency	<10	ng/mL
Insufficiency	10–30	ng/mL
Sufficiency	30–100	ng/mL
Toxicity	>100	ng/mL
Procalcitonin		
Lower risk	<0.5	ng/mL
Higher risk	>2.0	ng/mL
Parathormone	10–15	pg/mL
CRP quantitative	0–10	mg/L
CSF Biochemistry		
CSF sugar	50–80	mg/dL
CSF protein	15–45	mg/dL
CSF ADA-MTB	<10	IU/L
Pleural Biochemistry		
Fluid sugar	70–130	mg/dL
Fluid protein	2.5–3.5	g/dL
Pleural fluid for ADA-MTB	<30	IU/L
Urine 24-hour protein	0–150	mg/24 h
Protein/creatinine ratio		
Normal	<0.2	
Moderate	0.2–2.0	
Severe	>2.0	

Reference Range of Specific COVID-19 Biochemical Test

Biochemical parameter	Reference value	Units
LDH	200-400	U/L
CRP	0-10	mg/L
Ferritin	30-400	ng /L
Interleukin - 6	< 7.00	Pg/mL
Procalcitonin	< 0.5 – low risk > 2.0 –high risk	ng/mL
D -Dimer	< 500	ng/mL

Source: ROCHE Diagnostics.

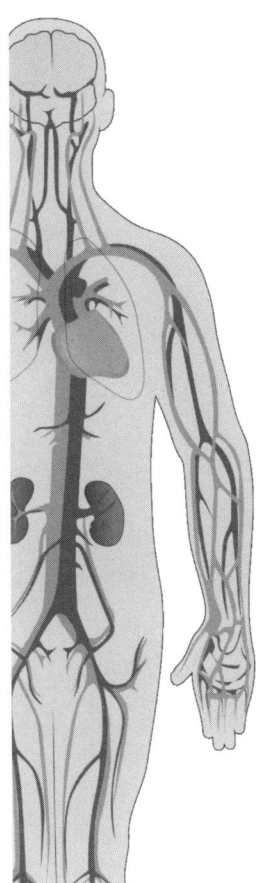

Main laboratory abnormalities observed in adult patients with unfavorable COVID-19 progression *(Modified 1-30)*		
Laboratory test	**Abnormalities**	**Potential clinical significance**
Complete blood count	Increased white blood cells Increase neutrophil count Decreased lymphocyte count Decreased platelet count	Bacterial (super) infection Bacterial (super)infection Decreased immunological response to the virus Consumption (disseminated) coagulopathy
Blood gases	Estimated modifications	Important in critical care management
Albumin	Decreased	Impairment of liver function
LDH	Increased	Pulmonary injury and/or widespread organ damage
ALT	Increased	Liver injury and/or organ damage
AST	Increased	Liver injury and/or organ damage
Total bilirubin	Increased	Liver injury
Creatinine	Increased	Kidney injury
Urea	Estimated increase	Kidney Injury
Cardic troponin	Increased	Cardiac injury
D-Dimer	Increased	Activation of blood coagulation and/or disseminated coagulopathy
Prothrombin time	Increased	Activation of blood coagulation and/or disseminated coagulopathy
Procalcitonin	Increased	Bacterial (super)infection
C-reative protein	Increased	Severe viral infection/viremia/viral sepsis
Ferritin	Increased	Severe inflammahon
Cytokines	Increased	Cytokine storm syndrome

Source: International Federation of Clinical chemistry and Laboratory Medicine.

UNIT 8

Practical Spots in Biochemistry

OUTLINE

8.1. Practical Spots in Biochemistry

Practical Spots in Biochemistry

8.1

SPOT-1
a. Identify the instrument.
b. Add a note on its principle and clinical applications.

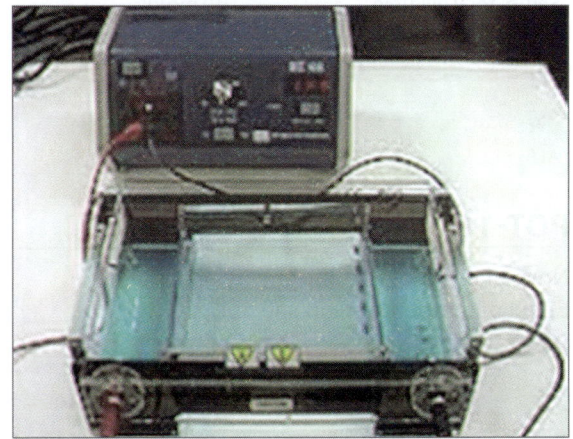

Fig. 8.1.1

SPOT-2

Fig. 8.1.2

a. Identify the given test.
b. Write down the constituents of the reagent used for the given test.

SPOT-3
Following are the values of oral glucose tolerance test performed on an individual. Write the probable diagnosis.

Time (in Hours)	Blood glucose (mg %)	Urine sugar Benedict's method
0	135	Blue
0.5	150	Blue
1.0	290	Yellow
1.5	268	Yellow
2.0	260	Yellow

SPOT-4
a. Identify the instrument.
b. Give its principle and clinical applications.

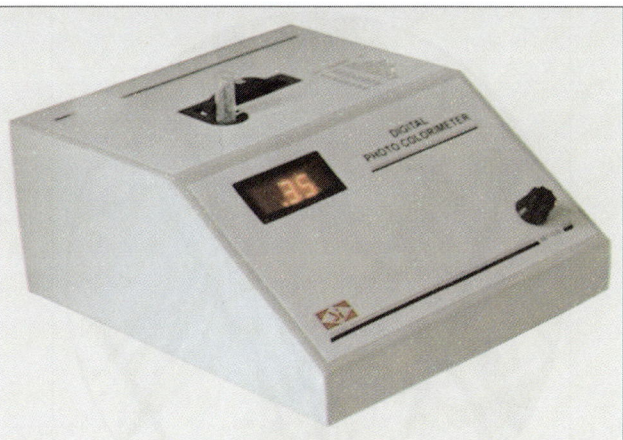

Fig. 8.1.3

SPOT-5
a. Define renal clearance.
b. Write the name of a fructan used to test glomerular function.

SPOT-6

a. Describe electrophoretic pattern of plasma proteins.
b. Which protein will increase in multiple myeloma and at what position?

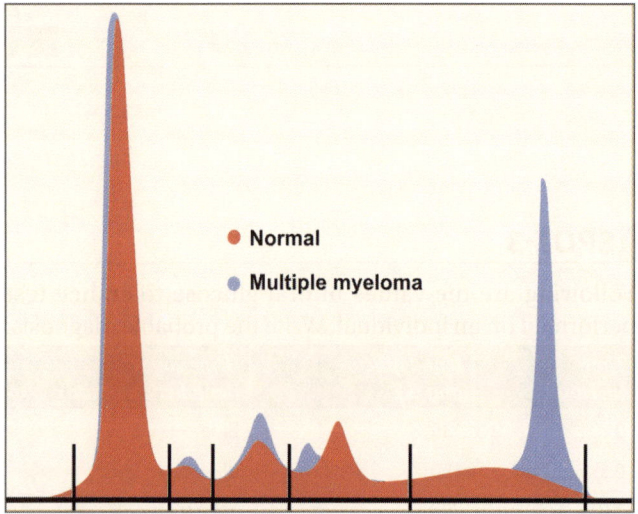

Fig. 8.1.4

SPOT-7

a. Identify microscopic view of the crystals.
   1. Identify the test.
   2. Name two substances which form these crystals.

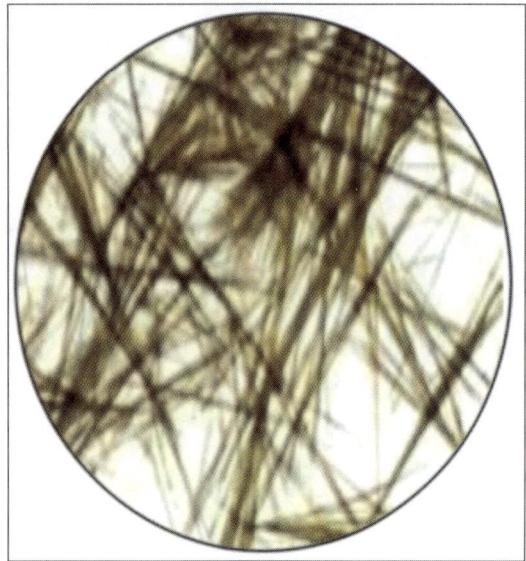

Fig. 8.1.5

SPOT-8

a. What is the normal urine output in 24 hours and its specific gravity?
b. Define polyuria and anuria.

SPOT-9

a. What is the full form of ELISA?
b. Give its principle and clinical applications.

SPOT-10

a. Following are some of the biochemical findings in a patient. Write the most probable diagnosis.
b. Justify why?

Serum total bilirubin	10 mg/dL
Serum direct bilirubin	8 mg/dL
ALT	52 IU/L
AST	43 IU/L
ALP	80 IU/L

SPOT-11

a. Normal range of specific gravity of urine.
b. Effect of temperature on specific gravity.

Fig. 8.1.6

SPOT-12

a. What is protein denaturation?
b. Give four examples of protein denaturing agents.

SPOT-13

a. What is PCR?
b. Steps of PCR.
c. Application of PCR.

Practical Spots in Biochemistry

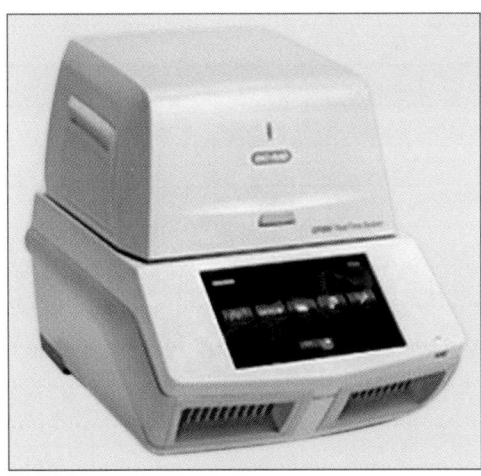

Fig. 8.1.7

SPOT-14
a. Sequence of enzyme required in DNA replication.
b. Enumerate DNA replication inhibitor.

SPOT-15
a. Define HbA1c and its clinical importance.
b. Which clinical conditions cause false interpretation of this investigation.

SPOT-16
a. Enumerate different methods of blood glucose estimation.
b. Which one is the best method among all and why?

SPOT-17
a. What are the cardinal symptoms of diabetes mellitus?
b. Describe diagnostic criteria of diabetes mellitus according to WHO.

SPOT-18
a. Classify immunoglobulin.
b. What is the function of different immunoglobulin, name of immunoglobulin found in body secretion?

SPOT-19
a. Which amino acid synthesizes nitric oxide (NO)?
b. Describe use of nitric oxide in treatment of different diseases.

SPOT-20
a. Give examples of radioisotope used in medicine.
b. Describe their clinical significance.

ANSWER TO SPOTS

UNIT 9

Competency-based Assessment for Practical Biochemistry

OUTLINE

9.1. Competency-based Assessment for Practical Biochemistry

Competency-based Assessment for Practical Biochemistry

9.1

						Section vs Practical			
Competency No.	Competency	I	F	IF	Σ IF	Section wise IF	Weight age (W)	Marks (80)	Certifications
Unit 1					143	08	0.55	4.47	
BI 11.1	Describe commonly used laboratory apparatus and equipments, good safe laboratory practice and waste disposal	2	2	4					NA
BI 11.19	Outline the basic principles involved in the functioning of instruments	2	2	4					NA
Unit 2					143	27	0.188	15.10	
BI 11.3	Describe the chemical components of normal urine	3	3	9					NA
BI 11.4	Perform urine analysis to estimate and determine normal and abnormal constituents	3	3	9					Yes
BI 11.20	Identify abnormal constituents in urine, interpret the findings and correlate with pathological state	3	3	9					NA
Unit 3					143	95	0.66	53.1	NA
BI 11.6	Describe the principles of colorimeter	3	3	9					NA
BI 11.7	Demonstrate the estimation of serum creatinine and creatinine clearance	3	3	9					Yes
BI 11.8	Demonstrate estimation of serum proteins, albumin and A:G ratio	3	3	9					Yes
BI 11.9	Demonstrate the estimation of serum total cholesterol and HDL cholesterol	3	3	9					NA
BI 11.10	Demonstrate the estimation of triglycerides	3	3	9					NA
BI 11.11	Demonstrate estimation of calcium and phosphorous	3	3	9					NA
BI 11.12	Demonstrate the estimation of serum bilirubin	3	3	9					Yes
BI 11.13	Demonstrate the estimation of SGOT/ SGPT	3	1	3					Yes
BI 11.14	Demonstrate the estimation of alkaline phosphates	3	1	3					NA

Unit 9: Competency-based Assessment for Practical Biochemistry

BI 11.17	Explain the basis and rationale of biochemical tests done in the following conditions: • Diabetes mellitus • Dyslipidemia • Myocardial infarction • Renal failure, gout • Proteinuria • Nephritic syndrome • Edema • Jaundice Liver disease, pancreatitis, acid base disorder Thyroid diseases	3	3	9					NA
BI11.18	Discuss the principles of spectrophotometer	2	1	2					NA
BI11.20	Demonstrate estimation of glucose, creatinine, urea and total protein in serum	3	3	9					NA
BI11.21	Calculate albumin: globulin (AG) ratio and its significance	3	3	9					Yes
Unit 4					143	06	0.04	3.35	
BI11.5	Describe screening of urine for inborn errors and describe the use of paper chromatography	2	1	2					NA
BI 11.16	Observe use of commonly used equipments/ techniques in biochemistry laboratory including: • pH meter • Paper chromatography of amino acid • Protein electrophoresis • TLC, PAGE • Electrolyte analysis by ISE • ABG analyzer • ELISA • Immunodiffusion • Autoanalyzer • Quality control • DNA isolation from blood/ tissue	2	2	4					NA
Unit 5					143	3	0.02	1.67	
BI 11.15	Describe screening of urine for inborn errors and describe the use of paper chromatography	3	1	3					NA
Appendices					143	1	0.006	0.55	
BI11.2	Describe the preparation of buffers and estimation of pH	1	1	1					NA

Footnote:
I: Importance (rated from 1-3)
F: Frequency (rated from 1-3)
IF: product of Importance and frequency
Σ IF= Sum of product of Importance and frequency

Section wise IF: Sum of product of importance and frequency in each section Weight age (W) : Section IF/ Σ IF
Marks: W x 80 (Maximum Marks)

Certification of Skill-based Competencies

This is to certify that Mr/ Ms. _____, Roll no. _____, University enrollment no. _____, Batch _____, has completed the following competencies in the Department of Biochemistry and has attained 'Shows How' level of procedural skill.

No.	Competency	Activity	Date completed:	Attempt at activity	Rating	Decision of faculty	Initial of faculty and date	Feedback Received
BI			dd-mm-yyyy	F: First or Only R: Repeat Re: Remedial	B: Below M: Meets E: Exceeds expectations OR Numerical Score	C: Completed R: Repeat Re: Remedial		Initial of learner
BI 11.4	Perform urine analysis to estimate and determine normal and abnormal constituents							
BI 11.7	Demonstrate the estimation of serum creatinine and creatinine clearance							
BI11.8	Demonstrate estimation of serum proteins, albumin and A:G ratio							
BI11.20	Identify abnormal constituents in urine, interpret the findings and correlate these with pathological states							
IM 11.12 (integrated with pathology and medicine	Perform and interpret a capillary blood glucose test							
IM 11.13 (integrated with pathology and medicine)	Perform and interpret a urinary ketone estimation with a dipstik							
PE 33.6 (integrated with pediatrics)	Perform and interpret urine dip-stick for sugar							

Signature of HOD

Record of Early Clinical Exposure

No.	Competency	Activity	Date completed:	Attempt at activity	Rating	Decision of faculty	Initial of faculty and date	Feedback Received
BI:			dd-mm-yyyy	F: First or Only R: Repeat Re: Remedial	B: Below M: Meets E: Exceeds expectations OR Numerical Score	C: Completed R: Repeat Re: Remedial		Initial of learner

Record of Self-directed Learning Sessions

No.	Competency	Activity	Date completed:	Attempt at activity	Rating	Decision of faculty	Initial of faculty and date	Feedback Received
BI:			dd-mm-yyyy	F: First or Only R: Repeat Re: Remedial	B: Below M: Meets E: Exceeds expectations OR Numerical Score	C: Completed R: Repeat Re: Remedial		Initial of learner

Record of Attitudes, Ethics and Communication (AETCOM)

No.	Competency	Activity	Date completed:	Attempt at activity	Rating	Decision of faculty	Initial of faculty and date	Feedback Received
BI:			dd-mm-yyyy	F: First or Only R: Repeat Re: Remedial	B: Below M: Meets E: Exceeds expectations OR Numerical Score	C: Completed R: Repeat Re: Remedial		Initial of learner

Record of Sports and Extracurricular Activities

S. No.	Activity	Date		Initial of faculty and date
		From	To	

Signature of Mentor

Final Summary

This is to certify that Mr/ Ms. _____, Roll no. _____, University enrollment no. _____, Batch _____, has completed the following modules and is recommended for appearing in Final University Examinations:

S. No.	Description	Dates From	Dates to	Attendance percentage	Status complete/ incomplete	Signature of teacher
1	Qualitative experiments and their clinical applications					
2	Quantitative experiments and their clinical interpretation					
3	Group experiment and Demonstrations					
4	Early clinical exposure					
5	Attitude, ethics and Communication					
6	Humanities					
7	Integrated biochemistry					
8	Sports and extracurricular					

Signature of HOD

Index

Page numbers followed by *f* refer to figure and *t* refer to table.

A

Achondroplasia 109
Acid-base
    balance, respiratory regulation of 198
    disorders 197, 198
    regulation 197
Acidosis 198
Acute pancreatitis, laboratory tests in 193
Addison's disease 145
Adrenal steroids, increased secretion of 145
Adsorption column chromatography 152, 153*f*
AETCOM module 211, 213
Affinity chromatography 153
Ag concentration 165*f*
Ag-Ab complex, plot of 165*f*
Agglutination 164
    indirect 167
    inhibition 168*f*
        assay 167
    reaction 165, 167
Albumin 57
Alcohol
    intake 126
    precipitation by 35
Alcoholism 128
Aldehyde test 38
Alkaline
    copper sulfate reduction methods 79
    phosphatase 92, 109
    tide 48
Alkaloidal reagents 35
    precipitation by 35
Alkalosis 198
Aluminum hydroxide 137
Aminotransferases 92, 105
Amiodarone 187
Ammonium ion, excretion of 198
Anemia, severe 109
Anorexia 145
Antibody
    concentration 165*f*
    detection 161
    excess 165*f*

Antigen
    antibody interaction 164
    detection 160
    effects of 165*f*
    excess 165*f*
Antithyroid agents 186
Anuria 57
Apoproteins 158
Arterial blood gas
    analysis 147, 148*t*
        applications of 148
    analyzer 147*f*
Arterial carbon dioxide, partial pressure of 148
Arterial oxygen, partial pressure of 148
Aspartate amino transferase 105
ATCOM sensitization 173, 184, 190, 195, 200
Atherogenic profile 124
Atherosclerosis 128
Atomic absorption spectrometry 135
Attitude, ethics and communication
    modules 203
    record of 240
Autoimmune
    disease 94
    thyroid disease 188
Automated analyzers 5
Autonomous nodule 187

B

Bacterial agglutination 167
Barfoed's test 18, 18*f*
Base excess 148
Beer's law 67, 75
Beer-Lambert's law 67
Bence Jones proteins 58, 60
Benedict's qualitative
    reagent, composition of 58
    test 58
Benedict's test 14, 15*f*, 21
    positive 58*f*
Benzidine 60
    test 60
Berthelot method 117

Bicarbonate 197
    reabsorption of 198
Bile flow, obstruction to 92
Bile pigments 59, 61, 92
Bile salt 59, 59*f*, 61, 92
Bilirubin 92, 102
    concentration
        of indirect 102
        of total 102
    glucuronide 59
Biochemical examination 49
Biochemistry, practical spots in 227, 229
Biological hazards 7
Biphosphate 197
Biuret 93
    method 93
    reagent 93
    test 28, 32, 39
Blood 175
    ammonia 92
Blood buffers 197
    types of 197
Blood glucose
    estimation of 78, 79
    self-monitoring of 79
Blood pigment
    Benzidine test 61
    test for 60
Blood sample 7
    collection of 78
Blood sugar
    clinical classification of 78
    fasting 78
    random 79
Body fluids 148*t*
Body language 205
Boiling water, solubility in 21
Bone
    carcinoma 109
    disease 94, 109
    isoenzyme 109
    neoplasm 135
Bouguer's law 75
Brain 182
    injury 145
Bromocresol green method 96

Bromosulphalein excretion test 92
Buffers and solutions,
    preparations of 221

C

Calcitonin 186
Calcitriol 113
Calcium 181
    role of 134
    test for 49, 50, 53, 53f
Calibration curves 76
Calomel electrode 142
Carbohydrate 22
    analysis of 13, 16, 17, 20, 23, 24
    metabolism 193
Carbonic acid 197
Carcinogenic chemical 60
Cardinal rules 7
Cation-exchange resin 88
C-context 205
Celiac diseases 94
Cells 181
    use of 75
Centrifuge 4, 5f
Cerebrospinal fluid 181
    analysis 181
        indications of 182
        characteristics of 182t
        composition of normal 181
Chemical composition 181
Chlorides 181
Cholecystokinin-stimulation test 192
Cholesterol 158
    ester 158
    oxidase 125
Chromatography 150
    ascending 152f
    descending 152f
    equipment 154f
Chylomicrons 158
Circulatory failure 145
Clinical biochemistry 1
    laboratory, development of 3
Clinical chemistry laboratory 3
Collagen diseases 94
Colorimeter 5, 5f, 68f
    components of 68f
    principle of 68f
Colorimetry 67
    principle of 67
Common equipment 3
Communication
    foundations of 215
    partnership and teamwork 207
    skills 215
Congenital disorders
    agammaglobulinemias 94

Conjugated bilirubin 59
Conjugated hyperbilirubinemia 103
Conjugated proteins 28
Control test 32
Conversational skill 215
Coronary artery disease 124, 128
COVID-19 225
    biochemical test 225
Creatinine
    clearance 113, 114, 116
    picrate 114
    test for 50, 53, 53f
Cretinism 109
Crigler-Najjar syndrome 103
Crystalline structure 21
Cuvettes 68

D

Dehydration 94
    severe 145
Deoxyribonucleic acid, quantification
    of 176
Desired lipid profile 130
Diabetes
    diagnostic criteria for 80
    mellitus 84, 126, 128
    untreated 94
Diabetic curve 87
Diabetic tests 223
Diacetyl monoxime method 117
Diarrhea 94
    severe 145
Dietary deficiency 137
Diode-array spectrophotometer 74
Direct bilirubin, concentration of 102
Disaccharides 13
Dispersion devices 73
Diuretics 145
Doctor-patient
    communication 216
    relationship 211
Double beam 74
Double immunodiffusion 166, 166f
Double pan balance 3, 4f
Drug 186
Dry chemistry technique 79
Dubin-Johnson syndrome 103
Duodenal contents, examination of 192
Dye-binding method 94
Dyslipidemia 132

E

Early clinical exposure, record of 238
Edema 96
Effective communication skills 205
Ehrlich's test 60

Electric charges 32
Electrode, combined 142
Electrolyte 223
    analyzer 145, 145f
    clinical interpretation of 145
    estimation of 144
Electronic balance 3, 4f
Electrophoresis 156, 166
    equipment 156f
    types of 156
ELISA
    competitive 161
    indirect 161
    principle of
        competitive 161f
        indirect 161f
        sandwich 161f
    sandwich 160
Emotion 206
Encephalitis 182
Endogenous water 144
Enediols 79
Enhanced glucose tolerance 87
Enzymatic methods 79, 121, 125
Enzyme defects 121
Enzyme-linked immunosorbent
    assay 160
Epitope and paratope 164
Erythropoietin 113
Exocrine 192
Exogenous hormones 192
Exogenous thyroid hormone 187
Exogenous water 144
Exploratory session 207
Extended glucose tolerance test 87
Extended lipid profile 130

F

Fallacies 14
Fanconi syndrome 137
Fasting glycemia, impaired 80
Filtration technique 153
First aid 8
Fluid 182
Fluorescent method 135
Fouchet's test 59
    reagent 61
Fractures, healing of 137
Frank diabetes 80
Free fatty acids 158
Fresh urine sample 49
Friedewald equation 130
Fructose 58
Fructosuria 58
Full saturation test 32
Fully automatic analyzers 5, 6f
Functional tissue, loss of 186

G

Galactose 58
Galactosuria 58
Gamma-glutamyl transpeptidase 92
Gas liquid chromatography 150, 152, 152f
Gel electrophoresis 156
Gel filtration chromatography 153
    principle of 153f
Gel permeation 153
Gestational diabetes mellitus 80, 90
Gilbert's disease 102
Glass electrode 142
Glucometer 79
Glucose 58, 181
    challenge test 87
    concentration of 79
    level, estimation of 81
    oxidase-peroxidase method 79
Glucose tolerance test 86
    curve 87f
    factors affecting 86
Glucosuria 58
Glutamate oxaloacetate
    transaminase 105
Glycated hemoglobin 86, 88
GOD-POD method 79
Gout 123
Graves' disease 187

H

Half saturation test 32
Hashimoto's thyroiditis 186, 188
Hay's sulfur test 59
    positive 59f
    reagents 61
Hazards
    chemical 7
    physical 7
HDL-cholesterol estimation 130
Heat coagulation test 28, 29, 29f, 57
Heavy metal ions 35
Heller's test 57, 60
Hemagglutination 167
Hematuria
    gross 60
    microscopic 60
Hemoconcentration 94
Hemodilution 94
Hemolysis 137
Hemolytic anemia 103
Hemolytic jaundice 126
Hepatic isoenzyme 109
Hepatic jaundice 111
Hepatobiliary obstruction 109
Hepatocellular damage 126
Hepatocellular jaundice 103
Hepatocellular parenchymal disease 135
Hexokinase method 79
High performance liquid
    chromatography 154
Hippuric acid test 92
Hopkin's Cole's test 38, 39
Hormones, production of 113
Hot air oven 4, 4f
Housekeeping 7
Hyperadrenalism 145
Hyperbilirubinemia
    neonatal 102
    unconjugated 102
Hypercalcemia 135
Hypercholesterolemia 126
Hyperglycemia 80
Hyperkalemia 145
Hyperparathyroidism 109, 135
    primary 137
Hyperphosphatemia 137
Hyperproteinemia 94
Hyperthyroidism 94, 126, 187
    primary 187
    secondary 187
Hyperuricemia 121
    causes of 121
Hypervitaminosis D 135
Hypobetalipoproteinemias 126
Hypocalcemia 135
Hypocholesterolemia 126
Hypoglycemia 80
Hypokalemia 145
Hypomagnesemia 135
Hyponatremia 145
Hypoparathyroidism 135, 137
Hypophosphatasia 109
Hypophosphatemia 137
Hypoproteinemia 93
Hypothyroidism 126, 128, 186
    primary 186, 187
    secondary 186, 187

I

Immunochemical techniques 164
Immunodiffusion 164, 166, 166f
Immunoelectrophoresis 158, 166, 166f
    modifications of 166
Impaired digestion 192
Impaired glucose
    regulation 80
    tolerance 80
Increased potassium levels 145
Indirect titration 134
Infection, chronic 94
Infertility profile 224
Infiltrative disease 186
Infrared spectrum characteristics 68t
Inherited autosomal defect 109
Inorganic phosphate 181
Inorganic phosphorus 137
Intestinal isoenzyme 109
Intestinal tract, disease of 109
Intravenous glucose tolerance test 87
Invert sugar 21
Iodine
    deficiency 186
    test 14, 15f
Ion-exchange chromatography 152
Ion-selective electrode 135
    advantages of 145
    principle of 145f
Iron profile 223
Ischemic heart disease 128
Isoelectric focusing 157
Isosthenuria 57

J

Jaffe's reaction 114
Jaffe's test 50
Jaundice 111
    obstructive 103, 126
    physiological 102

K

Kala-azar 94
Ketone bodies 59f, 60
    test for 59
Kidney
    function, assessment of 113
    loss through 145
Kidney function test 113, 114, 117, 121, 223
    profile 113
Kind and King method 109
Kinetic method 79
Knowledge, skill and performance 207
Kwashiorkor 93, 109

L

Laboratory hazards 7
Laboratory waste, disposal of 7
Lactation 135
Lactic acid 197
Lactose 58
Lactosuria 58
Lambert's law 67
Lambert-Beer's law 69
Layer chromatography 151
Lead acetate test 38

Lead sulphide test 39
Liebermann-burchard method 125
Light, source of 68
Lipid profile 124, 125, 130
Lipoproteins 158, 158f
    function 158
    separation 158
Listening skill 205, 215
Lithium 186
Liver
    cell damage, tests of 92
    cirrhosis 109
Liver function
    assessment of 92
    test 92, 93, 96, 102, 105, 109, 223
        profile 92
Low dietary intake 145
Low protein intake 93
Lowry method 94
Lundh test 192
Lymphocytes 181

M

Macroglobulinemias 94
Macropipettes 5
Maintaining trust 207
Malabsorption syndrome 126
Malloy-Evelyn method 102
Malnutrition 93
Mancini method 166
Meningitis 182
Metabolic acidosis 145, 198
    causes of 198
Metabolic alkalosis 198
    causes of 198
Microalbuminuria 57
Micropipettes 5
Microscopic examination 49
Milk-alkali syndrome 135
Millon Nasse's test 39
Millon's test 38
Mineral acid 35
Molarity 221
Molecular exclusion chromatography 153
Molecular weight, determination of 157
Molisch's test 13, 14f
Molybdenum blue 79
Monosaccharides 13
Multinodular goiter 187
Multiple myeloma 58, 94, 135
Multiple sclerosis 182
Myocardial infarction 132

N

Nephelometric method 94
Nephritis 135
Nephrosis 135
Nephrotic syndrome 93, 100, 126, 128
Neumann's test 38
Ninhydrin test 38, 39
Non-absorbable antacid, ingestion of 137
Nonalcoholic steatohepatitis 107
Noncarbohydrates 15
Nonsugar reducing substances 58

O

O-cresolphthalein complexone method 134
Odor 48, 57
Oliguria 57
Optics 73
Optimum Ag-Ab concentration 165f
Oral contraceptive pills 128
Oral glucose tolerance test 86, 223
Organic phosphorus 137
Ortho-toluidine method 79
Osazone test 21
Osazones under microscope, shape of 21f
Osteomalacia 135, 137
Ouchterlony method 166
Overhydration 94

P

Paget's disease 109
Pancreas 192
    indirect stimulation of 192
Pancreatic diseases 192
Pancreatic function tests 192
Pancreatic profile 192, 223
Pancreatic stimulation 192
Paper chromatography 151
Paper electrophoresis 156, 157f
Parathormone 137
Partition chromatography 151
Passive agglutination 167
Pentoses 58
Pentosuria 58
Peroxidase method 125
Personal behaviour 7
Personal habits 7
pH 148t
    measurement, method of 141
    meter 141, 141f
        components of 142
    urinary 57
Phenylalanine 38
Phosphate test 49, 50
Phospholipids 158
Phosphoric acid 197
Phosphorus, test for 53, 53f
Phosphotungstate 130
Photocell 68
Photoelectric colorimeter 68
Phototube 68
Pipettes 5, 5f
Placental isoenzyme 109
Plasma
    colloidal pressure 96
    electrophoresis of 158
    proteins 93
Plasticware 5
Poliomyelitis 182
Polyacrylamide 156
Polycythemia vera 135
Polysaccharides 13
Polyuria 57
Positive heat coagulation test 58f
Postpartum thyroiditis 187
Postprandial blood sugar 78, 79
Potassium, decreased levels of 145
Potentially toxic drugs 113
Practical biochemistry
    blueprint of 235
    competency-based assessment for 233, 235
Precipitation method 130
Precipitation reaction 35, 165
Precipitation test 35
Precision and accuracy 170
Pregnancy 128, 135
    test, principle of 168f
Primary cause 121
Protein 57, 158, 181
    analysis of 28, 30, 31, 33, 34, 36, 37, 40, 56, 41, 47
    bound calcium 134
    color reactions of 38, 39f
    derived 28
    error 96
    estimation, methods of 94
    heat coagulation test 60
    losing enteropathies 94
    migration 157f
    precipitation reactions of 32
    SDS complex 157
    separation technique for 97
    simple 28
    tests for 57
Prozone effect 167
Pyruvic acid 197

Q

Quality control 170, 173
    analysis of 171
    monitoring of 171
    program, monitoring of 171
Quantitative analysis 75

R

Radial immunodiffusion 166, 166*f*
Radioactive iodine uptake test 187, 188
Reactions, precipitins 165
Reducing sugars, test for 14, 58
Reitman and Frankel's method 105, 107
Renal cause 135
Renal damage
    extent of 113
    monitoring progression of 113
Renal failure 120, 137
Renal function
    impaired 115
    test profile 113
Renal glycosuria 87
Renal regulation 198
Renal tubular
    acidosis 145
    necrosis 137
Renin 113
Respiratory acidosis 198
    causes of 198
Respiratory alkalosis 198
    causes of 198
Rheumatoid arthritis 94
Rickets 135, 137
Rocket electrophoresis 166
Rocket immunoelectrophoresis 167*f*
Rothera's nitroprusside test 59
    reagents 60
Rothera's test, positive 59*f*
Rotor syndrome 103
Ruhemann's purple 38

S

Safety and quality 207
Sakaguchi test 38, 39
Salt-losing nephritis 145
Sample collection and handling 6
Sarcoidosis 94, 135
Saturation tests 32
Scurvy 109
Secretin stimulation test 192
Self-directed learning
    exercises 139
    sessions, record of 239

Seliwanoff's test 18, 18*f*, 21
Semiautomated analyzers 5
Serum 158
    aminotransferases, estimation of 105, 106, 108
    amylase, estimation of 192
    calcium, estimation of 134, 136
    deproteinization of 137
    dilution 167
    enzymes in 192
    globulin, estimation of 96
    HDL-cholesterol, estimation of 131
    lipase 192
    protein 92
    sample 81
    SGPT 92
    total cholesterol, estimation of 125
    total protein, estimation of 95
    uric acid levels, estimation of 122
Serum albumin 92
    decreased 96
    estimation of 96, 99, 100
    increased 96
Serum alkaline phosphatase 92
    estimation of 109, 110
Serum bilirubin 92
    estimation of 102, 104
Serum cholesterol
    level 92
        estimation of 127
    reduces 126
Serum creatinine 113
    estimation of 114, 116
Serum phosphorus 134
    estimation of 137, 138
Serum triglyceride 128
    levels, estimation of 129
Serum urea 113
    estimation of 117, 119
SGPT, estimation of 107
Shell of hydration 32
Shock 145
Single beam 74
Single pan balances 3
Skill-based competencies, certification of 237
Sodium
    bicarbonate 8
    hypobromite test 49, 53, 53*f*
        reagents 50
    levels, decreased 145
Soyaben meal test 53*f*
Specific gravity 48, 57
    correction of 57
Spectrophotometer 5, 6*f*, 72*f*

Spectrophotometry, principle of 72
Spinal tumor 182
Split beam 74
Sports and extracurricular activities, record of 241
S-strategy 206
Standard clearance 118
Standard protein solution 93
Starvation 93
Steatorrhea 135, 137
Stock biuret reagent 93
Subarachnoid hemorrhage 182
Sucrose, inversion test for 21
Sugar values 88
Sulfur powder 61

T

T3 test 187
Technical skill 215
Tetrabromo-m-cresol-sulfonephthalein 96
Three-tier defence 197
Thyroglobulin 188
Thyroid
    antibody test 187, 188
    bacterial infection of 186
    carcinoma 187
        anaplastic 187
        follicular 187
        papillary 187
    disorders 186
    function test 186, 187
    gland 186
    hormone synthesis 186
        defect in 186
    peroxidase 188
    profile 224
    regulation of 186, 187*f*
    scan 187, 188
    ultrasound 187, 188
    viral infection of 186
Thyroiditis, subacute 187
Thyroid-stimulating hormone 186
    test 187
Thyrotoxicosis 135
Thyrotropin-releasing hormone 186
Tissue 175
Titratable acids, excretion of 198
Total protein, estimation of 93
Total serum protein 92
Triacylglycerol 158
Triglycerides, estimation of 128
Tryptophan 38
Tuberculosis 182

Tumor markers 224
Turbidimetric method 94
Two-dimensional
　　　immunoelectrophoresis 167f
Tyrosine 38

U

Unknown protein, identification of 28
Urate, excess production of 121
Urea 50, 181
　　clearance 113, 117-119
　　test for 49, 53f
Urease test 49, 53, 53f
　　reagents 50
Uric acid 113, 121
　　decreased excretion of 121
　　estimation of 121
Uricase peroxidase 121
Urine 92
　　abnormal of pathological 60t
　　analysis 48, 56
　　　of pathological 63, 64
　　chemical composition of
　　　physiological 48
　　collection 49
　　enzymes in 192
　　formation of 113
　　normal 49
　　　constituents of 49
　　　of physiological 50t
　　pathological 58f, 59f
　　physical
　　　composition of physiological 48
　　　examination of normal 50t
　　　properties of normal 48
　　physiological 53f
　　sample, analysis of physiological 54, 55
Urinometer 48
Urobilinogen 92
　　Ehrlich's test 61
　　test for 60
Urochrome 48
Uterus, tumors of 109

V

Viral hepatitis 111
Vitamin D intoxication 137
Vomiting 145
von Gierke's disease 126

W

Waste product 113
Wasting disease 94, 126
Water
　　bath 4, 4f
　　homeostasis 144

X

Xanthoproteic test 38, 39